D069036-4

THE
VEGAN
TRAVEL
HANDBOOK

An inspiring and practical guide to vegan-friendly travel

CONTENTS

TROPICAL

ROAD TRIP

Opposite: Vegans can get stuck into a cashew-based cheeseboard at Beetroot Sauvage in Edinburgh. Below: Black rice and roasted vegetables. Bottom: Vegan fare at Bali's Yoga Barn.

Why travel vegan?

You're probably used to a limited menu when you go out to dinner, but you should by no means feel that a short list of choices extends to your travels. Fear not, vegan traveller. The world is your (plant-based) oyster.

Close your eyes and imagine the following scenario: you've been invited to a dinner party at the home of someone you barely know, and have no idea what's being served. You can do one of a few things – contact the host and explain your preferences, have a light bite to eat before heading to the party, or bring over a vegan dish with enough to share. Troubleshooting your meals in a world of omnivores is a skill you no doubt practise on a daily basis, and perhaps without knowing it you've been training your entire vegan life for plant-based travel and adventure. Or maybe you're just dipping your toe into vegan waters. Either way, get ready to copy and paste scenarios from your everyday into a deliciously foreign locale, and sprinkle in some language barriers and incredible sights for good measure. A new country is just

a dinner party at the home of a friend whom you haven't met yet.

Consider this book your primer for how to practise what you already do in your home town, on the global go. Nimble as a culinary ninja, you're used to dodging unseen animal-derived ingredients and assembling meals from what others have relegated to edible margins as garnishes and sides. And just as your career or favourite colour hasn't been dictated by your diet of choice, don't limit your potential travel destinations by their per capita offerings of what you prefer to scoop on to your plate when back in the comfort of your own home. You can be vegan anywhere, and a growing number of people worldwide certainly are. It's just a matter of a little extra preparation and a flexible attitude. A sense of humour helps, too.

Are you an adrenaline junkie who happens to prefer tofu?

A history buff with a thing for nuts? A beach bum who adores tempeh? A culture vulture taking a holiday from animal products? Choose a place that interests you first, and figure out what you'll eat there later. Of course, if you're a gourmand whose number one hobby is eating, there are plenty of destinations for you to choose from as well. Their locations might surprise you.

Opposite: Plant-powered salad and smoothies with burrito at Bali's Peloton Supershop. Above: Tables are set at Rosendals' Garden cafe in Stockholm.

Veganism around the world

The hippest trendsetters may be on a vegan diet now, but the lifestyle certainly isn't anything new. Just ask Buddha, Gandhi and maybe even Pythagoras.

It is infinitely easier to pick blueberries than hunt for wild boar, which is partly why vegetables, fruits, nuts and grains have been staples of the human diet since, well, forever. However, as we stopped hunting and gathering, and started farming crops and animals, the amount of space those plants occupied on our collective plates changed. Over the centuries, some cultures indulged a ravenous appetite for steak; others stuck with animal-free eats of the leafy variety.

When Hinduism developed in the Asian Indus Valley around 2000 BC, one of its principles was that followers practise ahimsa – non-violence towards all life forms, including animals. The religion didn't veto any foods outright, but many Hindus became lacto-vegetarians, eating dairy and foregoing meat, fish and eggs. To this day, India has one of the largest meat-abstaining populations in the world; roughly 25% of the country's citizens are vegetarians. Most restaurants there have separate vegetarian and non-vegetarian menus, even McDonald's, which serves a vegetarian Maharaja Mac.

Buddhism, too, adheres to *ahimsa*, with monks and some believers sticking to a strict vegetarian diet (while most other Buddhists reserve their all-veg eating for special occasions, such as religious celebrations). Some Buddhist communities – including the 1300-year-old Koyasan monastery complex on a mountaintop south of Osaka, Japan – open their doors to pilgrims and visitors who are welcome to sample the monastic lifestyle. The menu at Koyasan is strictly *shojin ryori*, classic (and vegan) Japanese Buddhist

cuisine. A typical meal consists of *ichi ju san sai*: one soup, three sides, pickles and rice. *Goma dofu* dressed with wasabi and soy sauce is a *shojin ryori* delicacy; these beige cubes aren't made from soya beans but from a combination of sesame paste and kudzu root.

Fast-forward a few centuries and modern communities have also embraced plant-based eating. The Rastafarian religion, developed in Jamaica in the 1930s, promotes a mostly vegetarian and often plant-based diet called Ital for health and purity reasons – and also steers clear of additives, chemicals and processing. 'Ital is vital', any Rastafarian will tell you, and their one-pot stews dressed with coconut milk and spicy Scotch bonnet peppers are designed to reinforce a connection to nature.

Meanwhile, across the Atlantic, The Vegan Society was founded in the UK in 1944 and is the oldest vegan organisation in the world. It coined the now-ubiquitous term for an entirely animal-absent diet (although other names were proposed, including benevore, allvega and vitan).

As vegans grew in number, animal rights organisations such as People for the Ethical Treatment of Animals (PETA) helped the community expand their protest to platforms beyond their plate. PETA was founded in 1980 in the US to lobby for animal rights, promote a vegan diet and help people shop cruelty-free.

Regardless of the roots of their inspiration, vegans have grown in number over the past few years and the diet is becoming more mainstream. Travel destinations are learning to cater to vegans, just as they would anyone else.

PLANNING

How expensive is travelling as a vegan?

As a long-standing vegan, you'll already have plenty of experience in weighing up options and choosing what's right for you. Recent converts will be quickly catching up. Be just as discerning when you're planning your travel budget, and you won't break the bank.

Google 'top vegan restaurants in X', and you'll probably find a mouth-watering list conveniently laid out for you by a vegan blogger or foodie website. The suggestions look delicious and it can be easy to get caught up in the once-in-a-lifetime temptation of eating a vegan *injera* platter in Ethiopia or seafood-free paella in Spain. Yes, you are much less likely to find such delicacies in your home town, but your budget will take a beating if you hit every spot selling vegan bubble tea, *pierogi* and *pho*. Your money may be more economically spent finding a vegan-friendly cooking class so that you can sample authentic local fare (and learn to cook some simpler meals for the rest of the trip).

On the flip side, if the vegan options at your destination are generic, it can be depressing to constantly spend money on basic noodles and simple salads. Prioritise how you want to spend your dining budget, both in relation to your other travel costs and in terms of which meals are worth the cash, within the amount you've earmarked for eating and drinking. If your accommodation doesn't include breakfast, for example, prepare an inexpensive meal of overnight oats yourself so that you can splurge on lunch and dinner.

Previous page: Making use of the idyllic pool at Jakes, Jamaica. Left: A selection of super salad, vegan burger and corn chips at The Yoga Barn in Bali. Right: Table service on a vegan tour of Tel Aviv, Israel.

WHAT SHOULD I PRIORITISE IN MY BUDGET?

ACCOMMODATION

Cucumbers are cheaper than caviar, right? By that logic, vegan travelling should be comparably inexpensive. Sadly, that's not quite the case. Sometimes being a vegan costs less, and sometimes it doesn't – when you travel away from home, the same is true. Between flights, accommodation and eating out, the big-spend items on your budget will be the same as for any other traveller.

Meals might be a little bit cheaper (since salads and grains tend to be less expensive than meat, seafood and dairy), but your food choices might also push you towards certain upgrades, such as a stay in a hotel with a refrigerated minibar instead of a lower-cost guesthouse. Just like anyone else planning a trip, decide in advance what to prioritise and where you're comfortable skimping a bit.

When choosing a place to stay, look for spots that equip their guestrooms with mini-fridges or offer access to a full-sized communal one. This might be a bit pricier upfront, but will enable you to store leftovers and some basic ingredients so that you can prepare simple meals for yourself. If you can swing it, access to a full kitchen is ideal.

SELF-CATERING

Being able to put together a few simple meals for yourself is more important for vegan travellers, since you might occasionally go hungry otherwise. Bring some basic equipment, in addition to non-perishable snacks (such as nuts and dried fruit). A utility knife, some food containers, utensils, and a couple of salt and pepper packets from your in-flight meal should suffice. Overnight oats, salad, nut butter sandwiches and cut fruit are easy to prepare anywhere.

OFTEN-OVERLOOKED COSTS

Vegan main courses can be scarce at some restaurants, which can force you to assemble a more expensive meal from a few side dishes and starters. Your stomach may not sense a difference, but your wallet might.

So instead of setting your smartphone to airplane mode for the duration of your trip, it might be better to pay for a data package so that you can look up nearby vegan-friendly restaurants on the go.

5 tips for saving money as a vegan traveller

4 EAT IN

You can shell out for a pale iceberg lettuce salad at a restaurant (when it's your only choice on the menu), or rustle up a more satisfying meal yourself. A few basic supplies – some veggies from the local market, a plant-based spread and a fresh loaf of bread – can go a long way.

1 BOOK A VEGAN IN-FLIGHT MEAL

If you forget to reserve an in-flight meal, you'll go hungry or need to buy pricey airport food instead of eating what you already paid for when you booked your plane tickets (if your airline provides in-flight catering, that is). Most airlines still let you reserve a vegan meal a few days before take-off.

2 TRAVEL ACCORDING TO THE SEASON

Peak travel season is usually dictated by a combination of holiday schedules and optimal weather. Vegans might choose a different time of year instead, calculating the best month to feast on fresh foreign produce. If these peak and produce seasons don't overlap, you're in for some savings.

3 DOWNLOAD SOME FREE APPS

Use free apps on your trip: V Cards: Vegan Abroad (www.vegan.cards) translates your dietary preferences into the local language; and Vanilla Bean Plant-Based Food (www.vanilla-bean.com) maps out vegan-friendly restaurants in the area.

5 DO YOUR HOMEWORK

Research vegan restaurants and local vegan dishes in advance, so that you can make informed dining choices on the fly, dictated as much by what you can eat as by how much you want to spend.

FROM LEFT: © ANNAPURNA MELLOR / LONELY PLANET, © ADRIENNE PITTS / LONELY PLANET

JANUARY–MARCH

• It's best to scale Argentina's Andean peaks between December and February, during the country's summer months (p48).

• No matter when you visit Vietnam (*below*), it will be raining somewhere. Your best bet is to go in early spring if you want to visit the northern mountains (p60).

• India can be unbearably sticky during the summer monsoon. Try to plan your trip for winter, when it's also *chikoo* (sapodilla) fruit season (p92).

• It's beach weather down under in eastern Australia from January to March, so expect crowds of sun-worshippers (p152).

Where to go when

When picking travel dates, you might find yourself balancing the best weather conditions and festival dates against the agricultural calendar. Here's an overview to help you decide how to plan your trip according to climates or your appetite.

APRIL–JUNE

• Go to Japan (*below*) in April, when the island's spring bounty is harvested. Look for pickled plums, cherry-blossom *mochi* (sweet, sticky rice) and fresh bamboo-shoot rice (p72).

• You'll probably still need some cold-weather clothes to visit eastern Europe in early spring, but pre-Easter Lenten fasting in orthodox areas makes it the best time to find meat-, dairy- and egg-free foods (p52).

• The mild South African autumn is in April and May, which coincides with AfrikaBurn – the continent's version of Burning Man festival (p122).

JULY–SEPTEMBER

• Summer is the best time to go cloudberry, juniper, blackcurrant and lingonberry picking in Nordic countries, while also avoiding frigid temperatures (p76).

• Visit Brazil's Pantanal, the world's largest tropical wetland, during dry season (July through September). Birdwatching is also best during these months (p108).

• Fully blooming cherry blossoms and the Vegan Vibes festival make New Zealand a great September destination (p118).

• Catch the last of surfing weather during dry season in Bali (*below*) in September, right after most of the tourists have left (p66).

OCTOBER–DECEMBER

• An excellent time for hiking (*opposite*), you'll also avoid the crowds by travelling to Taiwan in its shoulder season towards the end of the year (p56).

• Jamaica (*below*) is mild in November and December, and local okra, callaloo and chocho crops are still available (p136).

• It's still beach weather in Israel's autumn season, when prices are lower and persimmons and pomelos flood the markets (p96).

• The 40 days leading up to Christmas (7 January in the Ethiopian calendar) are meatless for members of the Ethiopian Orthodox church (p52).

Packing

Your luggage probably looks pretty standard from the outside. Tucked inside, though, should be a few key items that set you apart from most other travellers.

1 PACK PACKAGING

Your suitcase will carry your clothes but how will you transport vegan snacks, restaurant leftovers and tropical produce? A few resealable snack bags, snack bag clips and a plastic container or two can help you stockpile plant-based deliciousness from one meal to the next. What is abundant in one locale might be scarce at another – if you find plantain chips or roasted beans at one stop, squirrel some away for the following day in case you unexpectedly find yourself in a precinct of steaks and cheesy omelettes.

2 BRING A SANDWICH,

MINUS THE BREAD

Even with the most diligent pre-trip research, some places are just less vegan-friendly than others. But that doesn't have to hamper your holiday. Bread can be found in most places, and packing a small plastic jar of your favourite protein-rich spread (peanut or almond butter, perhaps tahini), for those last-resort instances when you can't find anything else to eat, helps take the pressure off. You'll better appreciate everything else your destination has to offer with a sandwich under your belt.

3 UTILITY KNIFE

Having a perfectly fresh pineapple but no knife to carve it open is cruel indeed. Don't let this happen to you, especially if you're travelling somewhere where fresh produce will be a culinary highlight. Pack a small utility knife but remember to put it in your suitcase – not your carry-on! Take a fork and spoon as well.

4   STOMACH-SETTLING MEDICINE

Fresh juices and pre-cut tropical fruit from street vendors are convenient and delicious. Sadly, they can also come at a gruelling gastrointestinal price. Take extra care when eating raw street foods but also bring a few over-the-counter cures in case you eat something that doesn't agree with you.

5   MILK FOR YOUR LATTE

Does a week of black coffee sound like no big deal, or like a singular brand of travel torture? If your morning latte is that thing you need to start your day right, you might consider taking a couple of miniature boxes of long-life non-dairy milk with you. It can be hit or miss whether coffee shops offer soy or almond, and at the very least you'll be able to sip a cappuccino at your hostel or hotel.

You'll need a check list of essentials on any trip, whether you're spreading nut butter (top), tasting coffee in Madrid (left) or buying fruit juice in Goa (above).

Local vegan communities

A growling tummy is the worst soundtrack for your adventure away from home. Food may or may not be the focus of your trip, but to make sure you're adequately fuelled to maximise your travel time, ask a local.

BEFORE YOU GO

Set yourself up for gastronomic success before your trip by connecting with local vegan communities online. Google vegan bloggers who are from your travel destination, since plant-based foods are literally their bread and (non-dairy) butter. If your search isn't turning up any bloggers, look for vegetarian or vegan societies – there may even be events or festivals happening during your travel dates.

Lots of locations, including Berlin, Cambodia and South Africa, have designated vegan Facebook groups. Search for the country or city you'll be staying in, plus 'vegan'. Vegan Travel is a near 30,000-strong Facebook group of plant-based food lovers who like to eat well abroad, and it's a great resource for finding people living in your destination (or who have recently travelled there).

Couchsurfing (www.couchsurfing.com) and Meetup (www.meetup.com) are also good ways to get some face time with local vegans. Check whether any Couchsurfing hosts are vegan (regardless of whether you choose to crash on their couch), or if any vegan Meetup events are happening while you're in town.

And if you're travelling solo, check Veg Travel Buddies (www.vegtravelbuddies.com) to see if any other vegans have similar travel plans and want to occasionally share a table.

You'll find hubs of veganism across the globe, from food trucks on America's East Coast (left) to Berlin's Veganes Sommerfest (below left) and culinary tours in Iran (below).

WHEN YOU'RE THERE

It's usually a major faux pas to ask your waiter where best to enjoy your next meal. As a vegan traveller? Go for it! At most destinations, all you need to do to ensure a satisfying succession of meals is identify one recommended vegan restaurant in advance and head there on your first day. The staff are often vegans themselves, or at least plugged in to the plant-based scene. As a tourist, don't be shy about asking a restaurant to kindly refer you to their competition – let your waiter or waitress, or the diners sitting at the next table, tell you where else they like to eat vegan-friendly meals in town.

Travelling with meat-eaters

Your lunches tend to be a combination of kale and quinoa, but your travel partner requires a daily dose of chicken. How do you ensure that you both leave the table sated and satisfied?

PLAN A SPECIAL MEAL (OR TWO) IN ADVANCE

You might have your heart set on going to an organic farm with vegan cooking classes, and your travel partner may be hankering for an all-you-can-eat steakhouse. You don't necessarily have the same decadent, final-night feast in mind. Make sure you don't end your trip without savouring those special meals by making reservations in advance, at least one for each of you.

DOUBLE UP ON SNACKS

If you're planning to self-cater a few meals, pack some snacks to share and others to eat separately. This might mean doubling up on things like sandwich fillings: a jar of almond butter for you, a can of tuna for your partner.

GET CREATIVE WITH YOUR MEAL PLAN

Research your destination's traditional cuisine, so that you can identify things to eat at a typical restaurant (instead of spending your trip eating at all-vegan places, as you might otherwise be tempted to do). Venture outside the menu's mains section and pinpoint a list of traditionally vegan side dishes, soups and starters that you can assemble into a full meal, if necessary, wherever you end up eating.

FIND VEGAN VERSIONS OF LOCAL FARE

Your partner might be reluctant to eat at too many vegan spots for fear of missing out on the local cuisine. It's a fair point. Look for places offering vegan versions of traditional dishes instead of international fare. Your partner may even compare the dishes and be surprised to discover which version is tastier.

TAKE TURNS

Accept that your appetites are going to be different sometimes, and rotate the designated restaurant-picker. You choose where to dine one night, your partner the next – ideally at places where you can both find at least one thing to eat. If your travel buddy has taken you to a particularly meaty spot, compromise by getting vegan dessert afterwards.

Opposite: You don't need to be vegan to enjoy the beetroot ravioli at Tallinn's Vegan Restoran V or the vibe at Bread and Chocolate on the Caymans. Left: Also on the islands, Saucha has tasty breads and spreads.

Language

As a vegan voyager, you should be fluent in saying more than just 'please', 'thank you' and 'where are the toilets?' in the local language. Learn how to describe what you do eat, and what you don't.

Every traveller stands to benefit from learning a few basic phrases before venturing somewhere where a different language is spoken, but this is especially crucial for vegans, who are advised to commit a few extra terms to memory.

Google Translate is one free way to learn a few key words in the local language, and has an audio button to help you practise pronunciation. It's always best to try to use spoken communication first, but if you get stuck there are easy ways to get scripts of explanatory text written in the local language.

A few app-based phrasebooks geared towards vegans – such as V Cards: Vegan Abroad and Vegan Passport, published by The Vegan Society (available in print and as an app) – spell out that you are vegan in numerous languages, and include a detailed list of what you do and don't eat (down to ingredients that normally only appear in small print, like casein and gelatin). Pull up the relevant card with text in the local language – or a picture – to show to waiters, market vendors or whoever is selling you food.

Be friendly and collaborative when communicating the things you've taken off your personal menu, and be willing to brainstorm a simple dish together with the person serving or preparing your food. But if you suspect that a combination of all these techniques still hasn't succeeded in getting your message across, consider saying that you have a food allergy (and be prepared to say that in the local language as well).

Words won't do justice to the vegan delights at Saucha in the Cayman Islands, (below) and Jakes in Jamaica (opposite).

YOUR NEW VEGAN VOCABULARY

Don't get lost in translation. The phrases that are universally helpful to know are: 'I am vegan' and 'I don't eat meat, seafood, poultry, eggs, dairy or honey.' For extra credit, learn the words (see p24–5) for ingredients that often lurk in vegan-sounding dishes, as well as culturally specific terms that come close to explaining your lifestyle. 'Vegan' is not a unanimously understood concept and doesn't always translate, but many cultures have their own ways of explaining a meat- and dairy-free diet.

Place your order for a dish that seems vegan. Then make extra sure there are no unwanted ingredients or garnishes:

JAPAN

VIETNAM

SPAIN

GERMANY

INDIA

le katsuobushi: Japanese for 'no dried bonito flakes', a kind of preserved fish commonly used as a condiment or for flavour in *dashi* broth.

Không nuóc mǎm: Vietnamese for 'no fish sauce', a condiment added to many dishes, even when no other animal-derived ingredients are included.

Sin jamón: Spanish for 'no ham', which is scattered as a topping on soups and salads.

Nein speck: German for 'no bacon', which is often diced and included in salads, noodles and other seemingly vegan dishes.

Nahin ghee: Hindi for 'no ghee', a clarified butter used in many Indian dishes that otherwise appear wholly plant-based.

Ask for the closest cultural equivalent to a vegan diet, in terms locals will understand:

JAMAICA

ETHIOPIA

RUSSIA

ISRAEL

CHINA

Ital: the Jamaican Rastafarian term for mostly plant-based, unprocessed foods.

Yetsom beyaynetu: the Amharic name for a platter of fasting-friendly Ethiopian plant-based foods, available in the country roughly 200 days a year.

Postno-ye menyu: the Lenten menus many Russian restaurants offer in the run-up to Easter, featuring vegan foods (though they may include some seafood).

Parve: a Hebrew term describing foods that have neither dairy nor milk (but may contain fish or eggs), in accordance with Jewish kosher laws.

Wo chī quán sù: how to explain in Chinese that you follow a Buddhist vegan diet (this may result in onion and garlic being omitted, too, so do specify).

Vegan travel quiz

Not sure where to start your planning? Take our vegan traveller quiz to figure out the type of trip that's right for you.

1. What's the first thing you put in your suitcase?

a) Hiking boots
b) Three novels
c) Your new cocktail dress
d) Sunscreen

2. What's the first thing you book, after your flight?

a) A hot-air balloon tour
b) A hot-stone massage at the plush hotel spa
c) Tickets to an award-winning musical
d) A cooking class on the beach

3. You're scrolling through the in-flight movies on the aeroplane. What do you pick?

a) *127 Hours*
b) *Eat Pray Love*
c) *The Da Vinci Code*
d) *The Beach*

4. What do you look for in the perfect hotel?

a) My ideal hotel isn't a hotel, it's a tree house
b) Sheets with an insanely high thread count
c) A juicy historic backstory
d) Ocean access within 20m

5. What do you hope to bring back from your trip?

a) Not too many bruises
b) A fully recharged battery
c) A handful of museum postcards
d) Extreme tan lines

6. When you travel, you make sure your wallet is stocked with:

a) A valid scuba-diving licence
b) An eye mask
c) My International Council of Museums membership card
d) That handy bottle opener that lays flat in my purse

7. You're cooking dinner for one. What do you make?

a) What I can from the fridge
b) A healthy soup that simmered all evening
c) A new risotto recipe
d) Takeaway from the local Caribbean restaurant

8. Your boss unexpectedly gives you the afternoon off. What do you do?

a) Head to a hiking trail
b) Binge on Netflix
c) Check out that new gallery exhibition
d) Stock up on cocktail ingredients

9. The perfect way to enjoy a sunset is:

a) By a campfire on a mountaintop
b) Soaking in a claw-foot bathtub under a huge skylight
c) From the rooftop of a temple
d) On the beach, cocktail in hand, sand between my toes

RESULTS THIS WAY

➸⟶

10. Your favourite way to work up a sweat is:

a) Rock climbing
b) In a sauna
c) Pounding on the treadmill while watching a good documentary
d) Swimming

11. It's your birthday. Your closest friends know you'll love what gift?

a) The newest GoPro
b) A yoga teacher-training workshop
c) Concert tickets
d) A quick-drying Turkish towel

12. Whose email newsletter are you most likely to open?

a) Your mountain-biking group
b) The local philharmonic orchestra's seasonal schedule
c) The local historical society's membership newsletter
d) Your favourite surf shop's sales

13. You'd consider adding a biography of which of the following people to your reading list?

a) Sir Edmund Hillary
b) Deepak Chopra
c) Julia Child
d) Bob Marley

14. It would be unbearable travelling with someone who wanted to spend all day:

a) Indoors
b) Running about non-stop
c) In the middle of nowhere
d) Somewhere that has the kinds of things I could see at home

15. What are the main items inside your bedroom wardrobe?

a) It's 50% camping equipment
b) Comfy clothes, yoga mat and journals I'll use some day
c) Outfits for evening occasions
d) My collection of bathing suits

16. Which of the following is most likely to be found in your sock drawer?

a) Woollen hiking socks
b) Japanese split-toe tabi socks
c) Frida Kahlo themed socks
d) Mismatched pairs; in a perfect world I'd always be wearing sandals

17. What's your guilty-pleasure splurge?

a) Vegan protein bars, perfect for camping trips or long treks
b) An organic sugar scrub
c) That pinot grigio from a historic Italian vineyard I visited
d) Fresh coconuts

18. A friend is visiting from out of town. Where do you offer to take them?

a) An urban rappelling workshop
b) A beer garden with live music
c) On a guided graffiti tour led by active street artists
d) The aquarium

Results

MOSTLY 'A'S – GREEN-TINTED ADVENTURER

Your food choices keep you close to the natural world, and on your travels you aim for the same. Thrilling ways to scale mountains, see incredible landscape views and sleep under the stars are your ideas of a good time (and if a few nicks and cuts are acquired in the process, so be it). In your day-to-day life you may work in an office, but when you're on holiday you like to be on the move and giving your muscles a good stretch. Locations such as Ethiopia (p52) and Vietnam (p60) will satisfy your need for adventure.

MOSTLY 'B'S – UNWIND, REST, REPEAT

You do enough hustling in your everyday life; when you're travelling you just want to chill. That doesn't necessarily mean you need a week in a hammock with a good book (although a little bit wouldn't hurt) but you do like to take things slow and pamper yourself. Maybe you'll see all the sights, maybe you won't. But you're definitely sleeping in, leisurely wandering serene surrounds and getting a massage at some point. Choose to visit Nordic nations (p76) or Bali (p66), that offer beautiful things to see at a laid-back pace.

MOSTLY 'C'S – CULTURE LOVER

A day spent bouncing from art museum to historic walking tour to concert hall is your idea of bliss. Some travellers visit foreign countries to see new landscapes; you're interested in people and the stories they have to tell. Why are the region's buildings shaped that way, who are the local heroes and what do people here do for fun? When travelling with a partner you're always the one reading and researching to get your bearings. Pick a locale with centuries' worth of stimulating tales, such as India (p92) or Spain (p102).

MOSTLY 'D'S – HERE COMES THE SUN

Like the ruby-red tomatoes you occasionally keep on your kitchen counter, you are an organism in need of sun – and lots of it. Cold weather can cause your mood to plummet, but a dose of sunshine quickly has you soaring to galactic levels. It's no coincidence that the foods you tend to heap on to your plate are sun-worshippers, too: pineapple, watermelon and coconut all rank high on your favourites list. Make sure your itinerary includes lots of sunbeams and book somewhere tropical like Malaysia (p142) or Jamaica (p136).

Staying healthy

You've probably encountered at least one person warning you about the nutritional detriments of an entirely plant-based diet, unaware that the majority of vegans are proactive about staying healthy. Follow the same rules when you're on the road. And if you're new to veganism, pay extra-close attention.

Below: Avocado and courgette tartar at Ibiza's Wild Beets. Opposite: Find meat-free food in Tel Aviv.

FEED YOUR BODY WHAT IT NEEDS

Appetite-curbing carbohydrates such as bread, rice and noodles are commonly found everywhere, as is fresh produce. You won't starve on your trip. But plant-based protein may be in short supply, depending on the location. To be on the safe side, bring a few protein bars or small bags of nuts with you and look for more of the same once you arrive at your destination. This doesn't have to be a drag, make it a fun task: look for variations that you wouldn't find at your local supermarket, like spice-glazed roasted chickpeas.

If protein powder is more your thing, consider bringing a small amount that you can mix into fruit juices or other drinks. And just as you would at home, be sure to stay hydrated.

AVOID WHAT YOUR BODY DOESN'T NEED

Some of the easiest-to-find vegan options happen to be less than stellar for your body. (We're looking at you, fried potatoes, jam on white bread with zero fibre and plain steamed rice.) It can be too easy to lean on these ever-present crutches, especially after a tiring day of travel when all you want is a no-fuss meal and an early night. In the long run, though, eating too many unhealthy foods will hamper your trip by giving you a) indigestion, b) not enough energy to hit the trail the following day, or c) both.

Make it straightforward for yourself to stay healthy by keeping nutritious snacks at hand, and get into the habit of visiting a supermarket or fresh produce market at least once on your trip. As an added benefit, shopping for food can provide an excellent insight into local life.

• **Raw street foods that are less than fresh:** Raw pre-cut fruit sold in plastic bags or on skewers is a common street snack but can also carry germs. Check how clean the street vendor's knife looks, or how long that fruit has been sitting in the sun.

• **Ice cubes:** If ordering a fruit smoothie in a place where you've been drinking strictly bottled water, make sure no ice is being blended in.

• **Raw vegetables:** Opt for cooked rather than raw vegetables in countries where the tap water isn't recommended for drinking. If preparing a raw salad yourself, wash vegetables with bottled or filtered water.

• **Not getting enough calories:** If you're in a country with limited vegan options, you might find yourself subsisting on a basic wholefoods diet of raw vegetables and fruits. Get enough of the other food groups via your stash of protein-rich snacks.

Resources

VEGAN TRAVEL BLOGS

• **The Nomadic Vegan**
Wendy Werneth has visited more than 100 countries, collating her destination-specific vegan tips on an interactive world map. *www.the nomadicvegan.com*

• **Vegan Travel**
An aggregate blog where vegan travellers share tips from recent adventures. Country-specific guides are divided into general blog posts and individual restaurant reviews. *www. vegantravel.com*

• **Burger Abroad**
The personal blog of full-time vegan traveller Amanda Burger includes illustrated destination tips covering an enviable number of cities. *www. burgerabroad.com*

• **HappyCow**
Veteran app mapping vegan and vegan-friendly restaurants and stores worldwide. *www.happycow.net*

• **V Cards: Vegan Abroad**
Free app that has a written script in more than 100 languages explaining that you're vegan and what foods you don't eat. Select the relevant card and show it to your waiter. *www.vegan.cards*

• **Vanilla Bean Plant-Based Food**
Free app mapping vegan, raw and gluten-free options, mainly in Europe and North America. *www.vanilla-bean.com*

• **Vegan Passport**
Also available as a printed booklet, this phrasebook compiled by The Vegan Society explains your dietary preferences in languages understood by 96% of the world's population. *www.vegansociety.com/shop/ books/lifestyle-books/vegan-passport*

ACCOMMODATION

• **Vegvisits**
This site helps you book a stay at the home of plant-based eaters, get recommendations for local restaurants and markets, or book a kitchen by the hour to cook for yourself. *www.vegvisits.com*

• **Veggie Hotels**
A website listing vegetarian and vegan hotels, B&Bs and guesthouses from more than 60 countries. *www.veggie-hotels.com*

Left: An exquisite vegan spread at the Chillhouse in Bali. Below: The island's Peloton Supershop is also great for a plant-based feast.

Vegan operators & planners

- **World Vegan Travel**
Activities led by locals, veganised versions of local foods and swish places to stay in Rwanda, Vietnam, Thailand and France. *www. worldvegantravel.com*

- **Vegan Trip Planner**
Enter your destination and travel dates, then let someone else do all the travel research to create your bespoke plan. Personalised itineraries start at $99 USD. *www.theveganword. com/trip-planner*

OTHER WAYS TO FIND VEGAN FOOD

- **Airbnb Experiences**
Have a tailor-made food experience and meet a local. *www. airbnb.com/s/experiences*

- **EatWith**
Share a meal with a handful of strangers in the home of a local based in one of more than 130 countries. Book in advance. *www.eatwith.com*

- **Hare Krishna Restaurants**
All of the religious centres in the movement's international network operate affordable restaurants. *http://centers. iskcondesiretree.com/restaurants*

THE VEGAN TRAVEL HANDBOOK

INSPIRATION

TOP 20 VEGAN-FRIENDLY CITIES

So you want a no-hassle city break? A relaxing trip where you won't be scrambling for vegan-friendly choices? Don't stress, these 20 cities are more than ready for you.

1 BERLIN, GERMANY

Frequently cited as the vegan capital of the world, Berlin's secret comes down to its thriving population of vegans: at least 80,000 and growing rapidly. This is a city where you'll find not just vegan restaurants but businesses thinking creatively about your everyday needs: animal-free butchers, vegan hotels, vegan bars and Schivelbeiner Strasse, a 'vegan avenue' featuring cruelty-free clothing stores.

2 LONDON, UK

There seem to be vegan options within easy range of every Tube stop in London: fancy Ethiopian, Caribbean, Asian or Japanese vegan cuisine? What about the world's first vegan chicken shop? Even Pizza Hut offers vegan cheese at its London outlets.

3 NEW YORK CITY, USA

Whether you're looking for vegan restaurant chains, fine dining, burgers, crepes or doughnuts, New York City wholeheartedly embraces your needs. Though the Upper West Side may have limited choices, other neighbourhoods have the city's go-get-'em attitude. Where else would you find a 'vegetable slaughterhouse'? Or a cafe with vegan sleeping pods for when you need a nap?

4 TEL AVIV, ISRAEL

Surprised to see Tel Aviv in the top five? Don't be, Israelis have the highest vegan population per capita in the world. Plant-based is a way of life here, so vegans are guaranteed to eat really well, with lots of locally grown produce. Think beyond the hummus and falafels and prepare to be wowed by recent innovations in vegan cuisine.

5 PORTLAND, OREGON, USA

The vegan dedication is strong in Portland. It has an entire mall that's vegan, hosts a vegan summer camp, sells vegan furniture, there's a vegan punk club, a critically acclaimed fine dining vegan restaurant, even a vegan strip club... you get the picture. Portland also claims to have the world's first all-vegan barbecue.

6 LOS ANGELES, CALIFORNIA, USA

The home of many vegan celebrities (Natalie Portman and Liam Hemsworth, among others), LA not only caters to the cool crowd but respects all those who consider the body a temple. If Los Angelenos aren't following a plant-friendly lifestyle, then they're certainly familiar with it. Musician Moby has a vegan restaurant here, with all profits going to animal rights organisations.

7 UBUD, BALI, INDONESIA

With no shortage of tofu and tempeh on menus, Bali is an easy choice for vegans, but Ubud is your ground zero. As the island's spiritual and cultural nexus, Ubud is all about feeding your soul. There are plentiful raw food options featuring fresh, local ingredients, not to mention a lot of vegan retreats in glorious tropical locations.

8 WARSAW, POLAND

A decade ago Warsaw held little appeal for vegans, but it's making up for lost time with close to 50 exclusively vegan restaurants. Enjoy Mexican fast food, sushi, Italian eateries, ice cream, burgers, French bakeries, Polish comfort food and more.

9 TORONTO, CANADA

With its very own vegan neighbourhood, nicknamed Vegandale – a dedicated block of vegan food, goods and services with its own touring festival (see p44) – Toronto should definitely be on your travel bucket list. More than half of Canadians, especially those aged under 35, regularly opt for vegan meat alternatives, and Toronto chefs are catering for plant-centric eating like never before.

10 MELBOURNE, AUSTRALIA

Australia's capital of cuisine always has its finger on the pulse, so the rise of vegan dining here is unsurprising. In fact, it's one of the fastest-growing vegan markets in the world: good news for travellers. You'll find vegan choices everywhere, but the popular inner-city suburb of Fitzroy is Melbourne's vegan heartland, thanks to more than 100 vegan-friendly options.

11   PARIS, FRANCE

Not often considered a top vegan-friendly destination, Paris has changed significantly in recent years to turn things around. Since April 2017, when the city's first vegan pastry shop, VG Pâtisserie, started selling gourmet delicacies, proving it was possible to meet French standards, the scene has exploded. You'll now find more than 60 totally vegan shops selling wares of all kinds.

12   TAIPEI, TAIWAN

Taipei's Buddhist history has contributed to it becoming one of Asia's most vegan-friendly hubs. You'll need to navigate the language barrier but, with more than 50 vegan restaurants to choose from, all sorts of fusion cuisines await.

13   PRAGUE, CZECH REPUBLIC

Following an increased interest in healthier lifestyles, this traditionally meat-heavy destination is suddenly sprouting (sorry) an abundance of vegan options. You can feast on vegan brunches or try Asian fusion, Venezuelan and Indian cuisines, along with vegan versions of Czech dishes.

14   KYOTO, JAPAN

Due to its fondness for fish, Japan may not be top of your list for vegan adventures, especially with *dashi* (fish broth) used as the master stock for so many dishes. But *shojin ryori* is a helpful phrase here: it's a traditional style of animal-free cooking in Buddhist temples. A modern vegan scene has also grown in Kyoto in recent years.

15 AMSTERDAM, THE NETHERLANDS

The Dutch might like their cheese but there are plenty of vegan offerings in Amsterdam, along with English-speaking locals who are savvy to plant-based requirements. There's a variety of cuisines – Vietnamese, Japanese, Caribbean – to try, along with restaurants such as Meatless District and The Dutch Weed Burger: the latter a lot more innocent than you're probably thinking (it's a 'House of Seaweed').

16 EDINBURGH, SCOTLAND

Edinburgh has become Scotland's vegan capital (only recently eclipsing Glasgow, if you're after a solid second Scottish choice), with incredible food, drink and products to enjoy among a very supportive vegan community. You can even get a vegan haggis samosa if you fancy it.

17 GHENT, BELGIUM

Claiming to be the first city in the world to introduce a meat-free day of the week (Veggie Day each Thursday, an initiative funded by the Flemish government in association with Ethical Vegetarian Alternatives), Ghent is one of the most easily navigable cities for vegans. The EVA has mapped out plant-friendly locales on Ghent's tourism website.

18 GOTHENBURG, SWEDEN

It may not be the capital of Sweden but Gothenburg is king when it comes to its vegan, eco-friendly and cruelty-free options. The old working-class district of Majorna, in particular, has become Gothenburg's vegan hub over the past decade, shedding its former rough reputation and thriving as a creative and popular neighbourhood, home to several crowd-funded vegan projects and cooperatives.

19 MILAN, ITALY

Is it all cheesy pasta and pizza in Milan? Yes and no, *miei amici*. Yes, there are small, family-run places that won't cater for vegetarians, let alone vegans. But if you know where to look, there are dedicated outlets offering the likes of vegan burgers with pink buns or Michelin-starred haute cuisine in ultra-stylish locations, not to mention super-fresh veggie produce.

20 STRASBOURG, FRANCE

Strasbourg has charming vegan cafes galore, including Vélicious, which doesn't hold back on promoting veganism on its restaurant walls alongside photos of celebrity vegans (it also runs a burger joint). There are a few vegan food trucks around and even a vegan/vegetarian pizzeria, which, combined with the cosmopolitan vibe of Alsace's capital, will make you feel right at home.

TOP 10
VEGAN TOURS

Want to outsource the admin and planning? Take a trip somewhere new with like-minded souls, knowing that all your vegan needs are well and truly taken care of? You can.

1. INTREPID VEGAN FOOD ADVENTURES

Intrepid Travel, known for ethically minded, small-group adventures, has recently updated its Real Food Adventures with tours for vegan gastronomes: eight-day trips to India, Thailand or Italy. The emphasis is on food – local experiences and hands-on cooking classes. It hopes to expand to all-vegan accommodation options. *www.intrepidtravel.com/uk/ vegan-food-adventures*

2. VEGAN EPICURE TRAVEL

Whether you're after a half-day excursion or a week-long global escape, Vegan Epicure Travel has plenty of choice. Would you like a cycling tour of Provence? What about a women's wellness retreat in Sri Lanka? A food tour in the French Riviera? A visit to a sanctuary of rescued animals in California? The hardest bit will be choosing. *www.veganepicuretravel.com*

3. SAMSARA VEGAN TRAVELS

The best thing about Samsara Vegan Travels is not just the range of destinations (Sweden, Myanmar [Burma], New Zealand, Vietnam, Nepal and Japan), but the freedom to customise your tour a little. Several opt-in activities and suggestions for free time mean you don't have to follow a set itinerary for the entire trip. *www.samsaravegantravels.com*

4. VEGAN SAFARI AFRICA

For animal-friendly eco-safaris in Botswana, look no further. Whether you want to splash out on five-star luxury accommodation or cheaper, authentic tenting experiences, Vegan Safari Africa has all budgets covered. There are a range of options for spotting wildlife in its natural habitat (*pictured*), from tiny white frogs in the Okavango to the infamous Big Five. *www.vegansafariafrica.com*

5. VEGAN RIVER CRUISES

Though some people may think a vegan cruise is an impossibility, Vegan River Cruises only uses new ships built in European shipyards that have met environmental guidelines and water protection regulations, and never burn heavy fuel. If that eases your conscience, choose from the Mekong Delta, Portugal, Myanmar (Burma) or even the Chilean fjords. *www.vegan-cruises.com*

6. VEGTHISCITY

Fancy a meat-free gastronomy tour of Singapore with a local? VegThisCity offers four-hour tours, from street-food strolls to sampling tea from all cultures. The tours are carefully curated by food lovers to highlight local communities and culture, as well as to give you a taste of the city's innovative vegan cuisine. *www.vegthiscity.com*

7. MELBOURNE VEGAN TOURS

You'll eat well in Melbourne, though all those laneways can make finding the vegan treats a bit challenging. Melbourne Vegan Tours (*pictured*) leads three-hour walks around Fitzroy, Collingwood and St Kilda, as well as the CBD. There's even a full-day Mornington Peninsula tour – the perfect excuse for a trip to the coast. *www.melbournevegantours.com*

8. UP NORWAY VEGAN TRAIL

A 13-day self-guided tour of Norway sounds spectacular at the best of times, but here you have the added bonus of a culinary vegan trip itinerary to help you along the way. Start with urban foraging in Oslo and finish with local vegan treats in Bergen. Car rental and accommodation with vegan dinners are part of the package. *www.upnorway.com/journey/the-up-norway-vegan-trail*

9. IRAN VEGAN TRAVEL

Iran's very first vegan-owned tour company has you tasting real Persian flavours cooked by vegan chefs, the way they were before meat was added in the last century. These are 12-day tours with a big focus on cultural immersion: spending quality time in Iranian family homes, attending Persian poetry readings and herbal distilling workshops, as well as exploring historical sites. *www.iranvegantravel.com*

10. TLVEG TOURS

These vegan walking tours (lasting about three hours) take you to vegan restaurants in Tel Aviv to meet chefs and sample street food. You might try an 'inside out cheesecake' while learning about the history of local veganism. *www.betelavivtours.com/eng/TLVEG_Tours*

TOP 10
VEGAN FOOD TRUCKS

Everyone loves the convenience of a food truck: quick cuisine at usually accessible prices made by friendly entrepreneurs. And the vegan food-truck scene is especially thriving.

1 GMONKEY, CONNECTICUT, USA

America's first vegan food truck has been lauded since its launch in 2010. Chef Mark Shadle has 20 years of gourmet vegan cooking experience and the truck runs on biodiesel fuel made from vegetable oil. Try the Zen Burger. *www.gmonkeymobile.com*

2 GOODDO, UDAIPUR, INDIA

The GoodDO food trucks are making it easier to be vegan in India. With a menu supplied by Good Dot, renowned for its 'mock meat' creations full of plant-based protein yet meat-like in texture, GoodDO's goodies include vegan meat tikkas, chilli wraps and vegan meat biryanis. *www.gooddot.in*; *@gudduzofficial*

3 TOTALLY AWESOME VEGAN FOOD TRUCK, PORTLAND, MAINE, USA

This '80s-inspired van is done out in a striking pink and purple 'synthwave' glow. Run by chef Tony DiPhillipo, TAVFT will make you smile with its Ruben U Been Waitin 4 (made with corned jackfruit pastrami), and nut- and gluten-free burgers. *www.totally awesomeveganfoodtruck.com*

4 CLUB MEXICANA, LONDON, UK

Craving piquant, hearty Mexican food? Club Mexicana serves pretty dishes stacked with handmade toppings and has a reputation for its juicy jackfruit burrito. Also on the menu are seitan, 'bacon' (tempeh) and 'tofish' (tofu) tacos, fully loaded nachos and, if you want to team a pint with your grub, vegan beers. *www.clubmexicana.com*

5 STAAZI & CO, ADELAIDE, AUSTRALIA

Bringing 100% plant-based food to Adelaide since 2017, the Staazi & Co food truck serves traditional Greek fare with a vegan twist. If you're craving the likes of *yiros*, moussaka or *spanakopita* you can get them here, with plant-based 'lamb' (TVP), 'mince' (soy and pea protein) and 'feta' (made with nuts), all prepared by passionate vegan Anastasia Lavrentiadis. *www.staaziandco.com.au*

6 THE CINNAMON SNAIL, NYC, USA

Breakfast burritos, French toast... it's all heavenly at this New York truck. You want a Beastmode Burger Deluxe? It's a grilled ancho chilli seitan burger with a stack of mouth-watering relishes and sides, on a pretzel bun. *www.cinnamonsnail.com*

7 CHEZ VEG'ANNE, STRASBOURG, FRANCE

Thanks to successful crowd funding from 253 backers, this vibrant green truck has been travelling around Strasbourg since 2016 and is known for vegan chef Anne Bavay's organic burgers on homemade bread, served with beetroot ketchup and sweet potato fries. There are also generous salads, quiches and desserts, but for a real sense of place go for the Veg'Alsace burger. *www.facebook.com/ chezveganne*

8 SOULGOOD, DALLAS, TEXAS, USA

At Soulgood, 5% of each purchase goes to the Cystic Fibrosis Foundation to honour the memory of Tyler, the youngest son of founder Cynthia Nevels. She began experimenting with vegan food when Tyler became terminally ill. It's all delicious, but if you're new to Dallas have the Cowboy Bowl. *www.eatsoulgood.com*

9 RUPERT'S STREET, LONDON, UK

The yellow Rupert's Street van changes its menu regularly, highlighting seasonal produce, which is good news for London growers and vegans. Depending on the time of year, you could try celeriac and potato cottage pie or beetroot and chickpea burgers. All packaging decomposes in less than 12 weeks. *www.rupertsstreet.com*

10 SHIMMY SHACK, DETROIT, MICHIGAN, USA

A vegan, gluten-free food truck, Shimmy Shack has served American comfort food since 2013. Its black bean and sweet potato chilli, and the milkshakes, have many fans. *www.shimmyshack.com*

TOP 10
VEGAN FESTIVALS

Imagine the fun of a festival in a new destination but with the added comfort of knowing you'll be among a supportive community of vegans. Wonder no more.

1. VEGANES SOMMERFEST, BERLIN, GERMANY

You want music, games, athletics and a fashion show as well as vegan food? Get yourself to Europe's biggest vegan festival in Berlin's Alexanderplatz in August. Best of all, it's free! There's also a tombola to spin on the way out, so you can leave with a gift from one of the sponsors. *www.veganes-sommerfest-berlin.de/en*

2. VEGANDALE, USA & CANADA (VARIOUS LOCATIONS)

The vegans of Chicago, Miami, Toronto, New York and Houston eagerly await this touring food and drink festival. The name stems from a dedicated vegan city block in Toronto (see p37). Expect 100% vegan food, craft brews, wine, spirits and clothing. *www.vegandalefest.com*

3. VEGGIEWORLD, EUROPE (VARIOUS LOCATIONS)

Europe's oldest vegan trade fair now exhibits in 14 cities, including Paris, Barcelona, Copenhagen, Munich, Brussels and Zurich. You'll find vegan products from hundreds of exhibitors, cooking demonstrations, talks, yoga classes and film previews. Mingle with fellow vegans at these two-day events. *www.veggieworld.de*

4. VEGFEST, LONDON & BRIGHTON, UK

Plump for London (October) or Brighton (March, though it's taking a break in 2020). Brighton's VegFest has been going since 2009 but London's attracts the biggest crowds (14,500 people in 2018). All stalls, marketing and speakers are vegan, and you can expect tastings, cookery demos, documentaries, even comedy. *www.vegfest.co.uk*

5. VEGAN STREET FAIR, LOS ANGELES, CALIFORNIA, USA

Los Angelenos have a choice of two Vegan Street Fairs. First there's the free, daytime, family-friendly street fair in March with its carnival-like atmosphere. Then there's the paid-entry, more intimate, night-time fair for those aged 21 and over in late August/early September. Both are held across two days/nights. *www.veganstreetfair.com*

6. VEGAN GOURMET FESTIVAL, NAGOYA, KYOTO & TOKYO, JAPAN

Launched in Nagoya and later expanding into Kyoto and Tokyo, this free festival is the place for gluten-free *ganmodoki* (fried tofu fritter with veg) burgers, vegan *takoyaki* (usually diced octopus balls but here made from *konnyaku*, a gelatinous-like potato), hemp beer and more. *www.vegefes.com*

7. VEGAN CAMP-OUT, UK

Join vegans from more than 30 countries at the world's largest camping festival in late August and early September. Going strong since 2016 (the location keeps changing to keep pace with demand), its entertainment includes music, talks, workshops, yoga and DJs. Solo travellers are welcome to join the Facebook group to make new mates beforehand. *www. vegancampout.co.uk*

8. HONG KONG VEG FEST

The people of Hong Kong celebrate

all things vegan at this one-day free carnival in October. Run by volunteers, it features more than 70 outlets offering vegan products, health checks, cooking demonstrations, cruelty-free clothing and more. There's even an animal costume stall and a place that rents reusable containers. *www.vegfest.hk*

9. VEGAN FASHION WEEK, LOS ANGELES, CALIFORNIA, USA

Debuting in February 2019 at the Los Angeles Natural History Museum, this four-day fashion festival founded by animal rights activist Emmanuelle Rienda was such a hit that designer applications immediately followed for the October 2019 runway show. Products are 100% vegan and, though the event is currently VIP-invitation only, it's worth checking out if you're a designer or fashion follower. *www.veganfashionweek.org*

10. AUCKLAND VEGAN FOOD FESTIVAL, NEW ZEALAND

This new festival, held in February, is a gathering of vegan food trucks, aka paradise for street-food lovers. With an all-ages, family-friendly crowd, live entertainment, a marketplace and warm, ethically minded hospitality, it's one to watch. *www.aucklandfoodtruckcollective. com/vegan-food-festival*

ADVENTURE

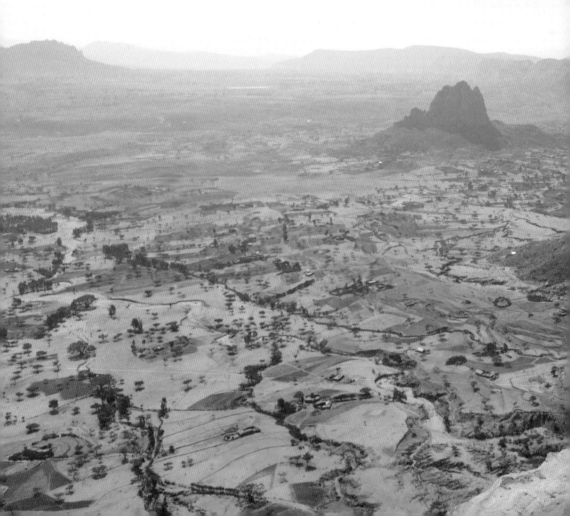

GREEN TIP

Argentine gastronomy does not yet value its greens. Travel with a high-quality green smoothie mix that doesn't require a blender, and perhaps some lightweight, easy-to-pack dried seaweed.

Practicalities

✈ Northern Patagonia (specifically El Bolsón), Buenos Aires (Palermo neighbourhood), Mendoza.

💼 Argentina comprises lush rainforest, high-altitude desert and windy Patagonia, so pack plenty of light layers as conditions change rapidly. Bring one 'nice' outfit for fashionable Buenos Aires.

📅 November is when purple jacaranda trees bloom in Buenos Aires, forming a stunning backdrop. March is a quiet time to visit Patagonia, as most summer tourists have left. Avoid Mendoza in late February and early March – Vendimia Festival makes booking difficult.

$ $ ▢

Argentina

Veganism is still a fairly new concept in Argentina, one of the world's most carnivorous countries. In general, the hot northern desert around Salta and lush jungle by Iguazú offer more fresh fruit and veg options.

Argentina is as picturesque as it is romantic. It's crossing the snow-capped Andes by Mendoza on horseback like a gaucho, wrapped in a wool poncho and sipping *yerba mate* to keep warm. It's learning to make pumpkin and walnut ravioli from kind sheep-herding Patagonian grandmothers with Italian heritage. It's dancing tango until dawn in sensual Buenos Aires, followed by breakfast at a cute vegetarian cafe.

Whether getting drenched by the Iguazú waterfalls, listening to the Perito Moreno glacier calve thunderously or hiking the seven-coloured mountains above Purmamarca, you'll benefit from an adventurous streak in this still-wild country. Vegans may have to search a bit among the red meat and cheese-slathered pizzas but, with perseverance, healthy food options can be found.

Most traditional Argentine dishes can be adapted but you'll often have to specify a meat-free version when eating out. Milanesa, usually made of chicken, can easily be made from aubergine. Empanadas can be stuffed full of veggies, lentils and fresh herbs. The north is much warmer and offers a larger range of fresh fruits and veggies at the market, but Patagonia (especially the El Bolsón, Lago Puelo and El Hoyo regions), stands out for its dedication to permaculture.

Previous page: En route to Ethiopia's Abuna Yemata Guh church. Opposite: The Falls of Iguazú National Park. Above: Don't miss the chance to tango in Buenos Aires.

Earthship Patagonia
A conscientious eco-lodging, the El Bolsón-based Earthship is also budget friendly. Stay in a yurt, private room or shared dorm. Complimentary breakfast includes hearty breads and homemade jams and granola with coconut milk. Every dinner is vegan and shared around a jovial communal table. *www.earthshippatagonia.com*

Palo Santo Hotel
With more than 900 plants covering the walls, plus vertical gardens and water fountains, this boutique hotel is a Buenos Aires oasis. There are free bikes to use and a Japanese restaurant on site. *www.palosantohotel.com*

Vines of Mendoza
There's no dietary request this vineyard-set hotel in the Valle de Uco doesn't meet. Each villa has a kitchen, and everything at its Siete Fuegos restaurant, including vegan options, is cooked over an open flame. *www.vinesresortandspa.com*

ESSENTIAL EXPERIENCES

Listen to local music
Attend a *peña* in Salta at La Casona del Molino, munching on vegetarian empanadas washed down with local *torrontés* white wine and soundtracked by folk music. *www.facebook.com/lacasonadelmolino*

Hang out with penguins
Head to Ushuaia to view penguins or a shipwreck with vegan-friendly Tierra Turismo. Later, scuba-dive the Beagle Channel with Ushuaia Divers, warming up after at historic cafe Almacén Ramos Generales. *www.tierraturismo.com*; *www.facebook.com/ushuaiadivers*; *www.ramosgeneralesush.com.ar*

Plan a well-fuelled trek
Stock up on healthy snacks at El Molino in El Bolsón, get some vegan takeout sandwiches at Café Alegria, then hike from Cajon del Azul to Glacier Hielo Azul. Finish with a veggie burger at Camping Base Hielo Azul. *www.facebook.com/el.molino.bolson*; *www.facebook.com/alegria.pasteleria*; *www.facebook.com/campobasehieloazul*

Follow the wine-tasting trail
Visit vineyards that produce vegan wine, such as Chakana, Andeluna, Zuccardi and Clos de los Siete. *www.chakanawines.com.ar*; *www.andeluna.com.ar*; *www.familiazuccardi.com*; *www.closdelossiete.com*

Take cookery lessons in the lap of luxury
It takes a plane, car, boat and deep pockets to get there, but learning how to cook over open flame on chef Francis Mallmann's Patagonian island is one epic vegan adventure. He may be famous for meat, but Mallmann is working on a vegan cookbook.

LEARNING THE LINGO

Useful phrases include 'Eu sou vegan' ('I am vegan') and 'Eu não como carne, peixe ou frango' ('I don't eat meat, fish or chicken'). Spell things out with 'Não como produtos de origem animal, incluindo lacticínios, ovos e mel' ('I don't eat products of animal origin, including dairy, eggs or honey') to make sure your preferences are met.

Left: Sampling the vegan wine of Andeluna Cellars. Below: Vegetable curry with coconut milk at Chirimoya in Salta.

LOCAL CUISINE

⟫⟶ Argentine gatherings centre around *asado*, a traditional barbecue usually of beef, lamb or goat, depending on the region. Bring along veggie burgers or stuffed red peppers to throw over the coals but expect some annoyed looks if you ask for your food to be cooked on a separate grill. Most restaurants will have white carb-heavy vegetarian foods such as pizza or pasta and will leave off the cheese on request. Salads in general are sad, often consisting of nothing more than iceberg lettuce and a couple of watery tomatoes. Shopping-wise, the local *dietetica* (health food store), stocking dried fruits, nuts, grains and supplements, should be easy to find in most cities.

BIO, BUENOS AIRES

The first certified organic restaurant in Argentina, this cosy cafe-style joint offers regular vegan cooking classes and a small marketplace. *www.biorestaurant.com.ar*

VERDE, BUENOS AIRES

Catering to vegetarian, vegan, gluten-free and raw diets, this Palermo restaurant is beloved by locals. *www.bsasverde.com/ alimentacion-inteligente*

DONNET, BUENOS AIRES

The menu in this Chacarita neighbourhood hideout is 100% vegan. It specialises in dishes based around mushrooms and also makes its own kefir.

CHIRIMOYA, SALTA

A world-class vegan treasure that makes pasta from courgettes, burgers from quinoa, and cannelloni from cornflour. It goes out of its way to make sure the beer and wine is all natural as well. *www.butterflypatagonia.com.ar*

GREEN TIP

Although vegan dishes reign on fasting days (prior to Leddet and Lent and every Wednesday and Friday), it's best to pre-order your meal at other times.

Practicalities

✈ The northern highland cities of Bahir Dar, Gonder, Aksum, Mek'ele and Lalibela.

💼 Shorts are reserved for children in Ethiopia, so it's probably best to leave yours at home. Pack a sun hat, hiking boots, loose-fitting casual wear and warm attire to don on the cold evenings.

📅 During the 40 days prior to Leddet (Ethiopian Christmas) on 6/7 January, and the 55 days preceding Lent, millions of Ethiopian Orthodox Christians fast on what is a strict vegan diet, which makes these periods an ideal time to travel to the country.

$ ☐ ☐

Ethiopia

Ethiopian food, like the country's landscape and unusual historical treasures, is unlike anything you'll encounter elsewhere. Due to regular animal-free fasting, vegan dishes are often the only thing on the menu.

The birthplace of an ancient civilisation and home to one of the earliest Christian empires on the planet, Ethiopia has proudly walked its own path for many millennia. You'll find that it still uses its own language, its own script, its own calendar and even its own way of keeping time. The country's cuisine is just as independent, enabling you to dine on a wealth of unique vegan dishes such as *shiro* (lightly spiced chickpea purée), *messer* (spiced lentil curry) and *tihlo* (barley balls), most of which are served atop the omnipresent *injera* (fermented, pancake-like bread). The nation also gave birth to the coffee bean, and no visit is complete without taking part in a traditional coffee ceremony.

However, your biggest discoveries will be away from the dining table. Northern Ethiopia's 'Historical Circuit' is truly a treasure trove: explore pre-Christian tombs beneath the towering stelae (obelisks) in Aksum; follow trails of incense into elegant rock-hewn churches that date back almost a thousand years in Lalibela; and stand in the shadows of towering 17th-century castles in Gonder. The circuit also offers access to the Simien Mountains, an unforgettable landscape for trekking and wildlife watching.

Opposite: A priest at one of Lalibela's Orthodox monasteries. Above: Cross-shaped Bet Giyorgis church, hewn from rock in Lalibela.

Hotel Maribela

A blend of contemporary style with touches of traditional fabrics will keep you lingering in Hotel Maribela's 15 light-filled rooms, which offer stunning views over the valleys from their ridge-topping perch in Lalibela. The equally scenic restaurant always aims to satisfy visiting vegans. *www.hotelmaribela.com*

Itegue Taitu Hotel

Locals and international visitors alike have long been drawn to Itegue Taitu Hotel's daily vegan buffet lunch. The establishment itself has a storied past – it was erected upon instructions from Empress Taitu in 1907, and the original building is steeped in its history. *www.taituhotel.com*

ESSENTIAL EXPERIENCES

Climb to a cliff-top church in Tigray

Work up an appetite by scaling the heights of Abuna Yemata Guh, a medieval rock-hewn church that sits spectacularly within a cliff face. Between the physical efforts required for the climb and the descent is the mental challenge of navigating the walk along a narrow ledge that hangs over a 200m drop en route to the church's entrance.

Learn how to cook like a local

Take part in a private cooking class in the home of an Ethiopian family. Enriching on both cultural and culinary fronts, sessions are arranged by Lodge du Chateau in Gonder. *www.lodgeduchateau.com*

Attend a coffee ceremony

A sign of both hospitality and respect, an invitation to a traditional coffee ceremony is something to embrace (your senses will thank you). The air is filled with not only the fragrances of roasting beans but also of burning incense and freshly cut grass scattered on the floor. Eventually, with the coffee's warmth radiating into your hands, savour your first of a minimum of three sweet cups.

Left: Join the locals and eat your fill at family-run Gonder restaurant Four Sisters.

ITEGUE TAITU HOTEL, ADDIS ABABA

The regal dining room of this Addis Ababa hotel is the perfect place to start your Ethiopian eating experience. The quality of the local fare is not only top-notch, but thanks to its daily (and great-value) vegan buffet lunch you can try an array of dishes to find your favourites. The restaurant's permaculture garden is also evidence of its aim for sustainability. *www.taituhotel.com*

FOUR SISTERS, GONDER

Further north, in the heart of Gonder on the Historical Circuit, join sisters Helen, Senait, Tena and Eden at their atmospheric and aptly named family restaurant. Take in traditional music and dance while you dine, and sign up for lessons on how to make *injera* or perform a coffee ceremony. *www. thefoursistersrestaurant.com*

LEARNING THE LINGO

A useful phrase is 'sega albellam' ('I don't eat meat'). Or just ask for your meal of choice, such as the traditional vegan fasting platter of yetsom beyaynetu, by adding alle. For example, 'yetsom beyaynetu alle?' ('Is there yetsom beyaynetu?')

LOCAL CUISINE

⟫→ The foundation of local cuisine is the vegan-friendly *injera*, a spongy sour pancake made from fermented grains of the teff plant. Mains are served as colourful piles on top of the *injera*, with the bread itself being used as a utensil to scoop up the meal (with your right hand only). Vegan dishes are ever-present on the weekly fasting days of Wednesday and Friday and during the extended Lent and Leddet fasts, but they are harder to find in smaller towns outside these times. Though *shiro* is the most available vegan dish, make sure you don't order *shiro be kibbeh* (*shiro* with clarified butter) or *bozena shiro* (with meat).

GREEN TIP

Bring your own flask to reduce plastic waste. There are free hot – and cold – water dispensers everywhere, in banks, museums, on trains and in all stations.

Taiwan

With its Buddhist history and love of tofu, Taiwan gets veganism. Wherever you venture, you'll find tropical fruits, tofu and veggie stir-fries, bean-based drinks and even desserts concocted from flowers.

Subtropical Taiwan is a below-the-radar adventure destination, great for all age groups and budgets. There are exhilarating water sports along its wind-whipped rugged coastline, a challenging 1200km cycling route that rings the main island, forests of giant trees, and the highest density of mountains in the world, topped by 3952m Yushan, the tallest peak in East Asia. You might also stumble across a secret temple or two in the misty hills.

Marvellously, it's also East Asia's vegan capital. Due to strong culinary influences from Japan, China and Southeast Asia, flavours are extremely varied and creative. Local favourites include mock-meat dumplings, creamy sesame paste noodles and three-cup tofu – a take on Taiwan's tangy *sanbeiji* (three-cup chicken) seasoned with sesame oil, rice wine and soy sauce. Traditional desserts are naturally vegan – silken tofu pudding, shaved ice and the ubiquitous bubble tea (ask for no honey or milk). Then there are the fruits – red dragon fruit, giant avocados, moreish custard apples and lychees – which can be packed on hikes or enjoyed in liquid form from a zillion fruit-juice stalls.

Opposite: Communal t'ai chi lessons in Tainan. Above: All prepped for a cooking class at Karenko Kitchen in Hualien.

Loving Hut Paradise Island

The owner of Paradise Island B&B whips up a different breakfast for her guests every day: think luffa vermicelli, mango toast, vegan omelette and fruit smoothies. Doubles and family-friendly quadruple rooms are well priced and the Penghu island location promises first-rate windsurfing, kitesurfing, snorkelling, diving and sea kayaking. *www.facebook.com/ paradiseisland88*

Sunnylands Farm B&B

This modern farmhouse homestay in the surfing hotspot of Yilan provides a meat-free countryside retreat, with the option to help out with the seasonal harvests of sunflowers, tomatoes, cucumbers, cauliflowers, guavas, bananas, sweet potatoes and lotus seeds – or just enjoy the fruits. *www. sunnylands.com.tw*

ESSENTIAL EXPERIENCES

Take a traditional cooking class

Experience Taiwan's unique indigenous tribal culture with a cooking class at Karenko Kitchen. Surprisingly, many dishes are vegan, including sticky rice wrapped in shell ginger leaves, crispy seaweed salad (plucked from fields and not the sea!) and fried wild vegetables. The school is located in Hualien, the jumping-off point for Taroko Gorge adventures. *www.karenko.com.tw*

Meditate in the mountains

The real roots of Taiwan's vegan culture lie in its Buddhist beliefs. Just north of Taipei, in rural Jinshan District, Dharma Drum Mountain offers overnight meditation retreats for a token fee. Dine with monks at this Zen Buddhist university complex tucked into the hills, then fall asleep to the sound of cicadas. *www.dharmadrum.org*

Taiwan Adventure Outings

The environmentally conscious and vegan-friendly TAO offers a menu of high adventures from night-time snake spotting to wild river tracing and mountain hiking across Taiwan's forests, mountain ranges and coastlines. *www.taoutings.com*

Eat your way around a night market

Taiwan's trademark night markets are found in even the smallest and remotest towns. Vegan favourites include grilled king oyster mushrooms, fried sweet potato balls (for calorie seekers), tropical fruits whole, diced or juiced, and – the ultimate cool-off in summer – mountains of shaved ice topped with sweetened beans, jellies and fresh fruit chunks.

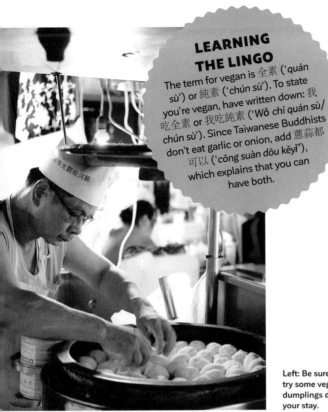

Left: Be sure to try some vegan dumplings during your stay.

LOCAL CUISINE

➤➤➤ Big cities abound with affordable vegan restaurants offering mouth-watering global menus: Japanese, Thai and Italian are all done supremely well. Budget-priced Buddhist-owned vegetarian buffets can be found all over Taiwan. Load your tray with veggies and mock meats made with soy or taro and pair with a bowl of purple five-grain rice. Be careful, some dishes contain egg or dairy. Convenience stores usually stock hot baked sweet potatoes, tofu rice pouches and bean- or nut-based milks. Vegan-friendly snack foods and packaged bread will be marked with the 全素/純素 label.

Top Restaurants

VEGE CREEK, TAIPEI & TAICHUNG

In many mall food halls in Taipei and Taichung, Vege Creek invites you to hand-pick ingredients – greens, pumpkin, aubergine, tofu skin, mushrooms – and select a noodle type. You'll be served with a filling, vegan version of the popular Taiwanese soy-sauce-spiced dish, *lǔwèi* (滷味). *www.facebook.com/ VEGECREEK*

HOSHINA UDON, TAIPEI

This vegan Japanese restaurant dishes up bouncy udon noodles, flavoured with sesame, kimchi, seaweed strips, tea or shredded vegetables. Watch the chefs making the noodles by hand. *www.hoshina.com.tw*

SOYPRESSO, TAIPEI

This hole-in-the-wall weaves magic with soy, serving bean-based silken puddings, yogurts, heavenly ice cream and bottled milk. *www.soy presso.com.tw*

GREEN TIP

Most dishes in Vietnam are seasoned with *nước mắm* (fish sauce) and even vegetable-only servings may include at least a dash. Practise saying '*không nước mắm*' (without fish sauce) with the proper rising and falling tones.

Practicalities

✈ Hanoi, Hue, Hoi An, Dalat, Ho Chi Minh City.

🧳 Bring warm-weather rain gear, especially if you'll be travelling during May to October. Warm clothing is a must if your itinerary includes northern Vietnam from December to March.

📅 Vietnam is a great place for tropical fruits, including durian, longan, lychee, mangosteen and rambutan, all in season June to August. Travel and accommodation rates leap in July and August and during Vietnamese New Year, which falls in January or February.

$ ▢ ▢

Vietnam

Exploring Vietnam's mountains, jungles and palm-fringed coastline invigorates body and soul, while its refined cuisine thrills the taste buds. Thanks to age-old Buddhist traditions, vegan fare is easy to find.

From the trekking trails of Sapa, high in the misty mountains, it's a 2000km drive to the white sands of Phu Quoc Island. In between – in addition to exploring dynamic cities – you can sail among the limestone pillars of Halong Bay, explore the vast karst caves of Phong Nha-Ke Bang National Park, kitesurf at Mui Ne and spot wild gibbons in the lowland rainforests of Cat Tien.

All around the country, in an array of regional variations, you can feast on Vietnamese cuisine. It's legendary for pairing vibrant flavours – accented by fermented sauces, syrupy dips, lime juice and hot chillies – with spectacular textures, the crunch often coming from fresh herbs, leafy greens and sprouts. Many dishes are based on animal products but owing to the strictures of Mahayana Buddhism, which requires that monks and nuns follow a vegan (or, some say, lacto-vegetarian) diet, classic Vietnamese dishes are widely available in vegan versions. Dishes to look out for include *rau muống xào tỏi* (morning glory sautéed with garlic), *cà tím mỡ hành* (grilled aubergine simmered with green onion) and *bánh xèo chay* (thin, savoury pancakes topped with mushrooms and bean sprouts).

Opposite: Street-side rice-cake vendors are a common sight in Vietnam. Above: Buy fresh dragon fruit at wet markets in Hanoi.

DREAM DIGS

Sofitel Legend Métropole Hotel

Opened in Hanoi in 1901, the venerable Métropole retains much of its French-era ambience. Graham Greene wrote part of *The Quiet American* here, and when Joan Baez visited in 1972 she took refuge – and sang – in a bomb shelter, rediscovered in 2011 (tours available). The hotel's restaurant, Le Beaulieu, serves superb vegan haute cuisine. *www.grandluxuryhotels.com*

CPG Vegan Trips

Colleen Patrick-Goudreau, author of several acclaimed books on vegan cooking and host of the podcast *Food for Thought*, runs vegan trips to Vietnam that emphasise ethical travel, local cultures, wildlife and, of course, out-of-this-world vegan cuisine. *www.cpgvegantrips.com*

ESSENTIAL EXPERIENCES

Slurp soup for breakfast

Once you get used to having *pho* (rice noodle soup) for breakfast, you'll wonder how you ever started the day without it! This aromatic Vietnamese staple, often garnished with bean sprouts, coriander, Thai basil and a wedge of lime, is traditionally made by simmering beef bones for hours. Fortunately, tasty vegan versions (*pho chay*) can often be found at and around Buddhist temples.

Head to the wet market

This is where Vietnam's freshest fruits, veggies and leafy greens are sold. Stalls (or, in the Mekong Delta's floating markets, boats) piled with colourful produce open early each morning.

Discover street eats in Ho Chi Minh City

Australian- and Vietnamese-run outfit Saigon Street Eats leads culinary tours of HCMC's backstreets, taking you to places you'd probably never be able to find without a guide. Highlights include street vendors and local wet markets. Bespoke vegan private tours can also be arranged. *www.saigonstreeteats.com*

Cook with a local

A class at the Green Bamboo Cooking School in Hoi An starts with a morning visit to the wet market to pick up fresh ingredients. Then chef Van shows participants – no more than 10 at a time – how to prepare authentic Vietnamese dishes, with more than a dozen vegan options. *www.greenbamboo-hoian.com*

LEARNING THE LINGO

Everyone will understand the concept of ăn chay (vegetarian eating). If clarification is needed, say 'tôi ăn chay' ('I eat vegetarian food') or ask that your meal be không (without) things such as trứng (egg), sữa (milk), thịt (meat) and cá (fish).

Left: Pull up a stool for an easy al fresco lunch at a Vietnamese market.

HUM VEGETARIAN, HO CHI MINH CITY

In a serene garden beside small lotus ponds, or indoors amid relaxing decor, diners at Hum feast on beautifully presented vegetarian and vegan cuisine made with natural ingredients. The menu includes salads featuring pomelo, banana blossom or winged beans, bitter gourd soup, and green and red curries. *www.humvietnam.com*

LIEN HOA, HUE

Imperial and vegan culinary traditions intersect in Vietnam's one-time royal capital, where this down-to-earth vegan restaurant is a favourite with Buddhist monks. Remarkably inexpensive highlights include fried spring rolls, aubergine with ginger, noodle dishes, crispy jackfruit and local beer. Staff don't speak English but the menu has rudimentary translations.

LOCAL CUISINE

⟫→ All Vietnamese cities and many smaller destinations have 100% vegan cafes and restaurants where meals are as inexpensive as they are delicious. Even restaurants that cater to an omnivorous clientele can almost always rustle up something vegan, though beware of being ambushed by nước mắm (fish sauce). Vegan food is easiest to find on the first and 15th days of the lunar month – when the moon is either new or full – when many Vietnamese Buddhists refrain from consuming animal products, and even many non-veggie restaurants and food stalls switch to vegan menus. The rest of the time, vegan eateries can often be found in the immediate vicinity of Buddhist temples.

RELAXATION

Practicalities

✈ South Bali (especially Seminyak and Canggu), Ubud, Amed, Singaraja/Lovina.

💼 Opt for beachy casual, but with modest long attire for visiting temples. Make sure you bring a rain jacket and sturdy shoes for any inland hiking.

📅 Bali's tropical climate means ultra-fresh ingredients year-round. Summer (July and August) is especially sunny and dry but also very busy. The shoulder season (April–June, September) is a bit more moderate in terms of both weather and crowds.

$ $ $

Bali

In Bali, modern green lifestyles meet ancient Indonesian culture. Start your leisurely day here enjoying raw banana pancakes and coconut kefir at a trendy vegan cafe, and end it by dining on tempeh at a food stall.

For decades, Bali has drawn travellers seeking something bigger than themselves. They find it in the volcanic island's temples, in its holy mountains, in the delicate flower petal offerings placed on every doorstep. They find it swimming next to sea turtles in the warm ocean, in watching the sun rise from black-sand beaches, in hiking through the emerald patchwork of rice paddies and the jungles resounding with monkey howls. And they find it in the island's growing wellness community, offering everything from juice cleanses and Ayurvedic massage to Balinese healing rituals and silent retreats.

For vegan travellers, the island is a double paradise. Forget scrutinising menus wondering if there's anything 'safe' to eat. In Bali, you can choose from dozens of international-style vegan cafes – think spirulina-laced smoothies, sprouted hummus, cashew cheese pizza – or just go local as Balinese cuisine is pretty vegan-friendly. Tofu and tempeh are both staples, dairy is rare and markets spill over with juicy tropical fruit including mangosteen, custard apple, jackfruit and pineapple. And with so many vegan visitors to cater for, most restaurateurs are quite familiar with your needs and are happy to accommodate them.

Previous page: The path to Ubud's Yoga Barn; find your Zen on a Bali beach. Opposite: Peloton serves up playful breakfasts in Canggu. Above: Canggu's Chillhouse eco-hotel.

Puri Ganesha Homes by the Beach

These thatch-roof luxury villas, with antique daybeds overlooking tropical gardens, are in a quiet corner of northwest Bali. The chef-owner, author of a raw vegan cookbook, fetes her guests with vegan meals incorporating local ingredients. *www.puriganesha.com*

The Chillhouse

Vegan aspiring surfers can sign up for a surf 'n' stay package at this cosy eco-hotel in Canggu. Fuel up on cassava and fresh fruit, stroll over to the beach for your lessons, then return for a free yoga class and communal dinner – food is all vegetarian, with many vegan options. *www.thechillhouse.com*

ESSENTIAL EXPERIENCES

Find your spiritual side

Wake in the cool Balinese morning and make your way to Ubud's Yoga Barn for an hour of sun salutations in the airy circular studio. The Barn is a hub for all manner of spiritual practices, with classes including qi gong, astrology and Tibetan bowl meditation. After your session, head to the cafe for an array of vegan choices, from cashew cheese lasagne to kale super-salads. *www.theyogabarn.com*

Cook and chow Balinese-style

In Seminyak, the Kin Vegan Cooking Class will teach you to make drool-worthy vegan Balinese dishes such as tempeh curry, tofu fritters with peanut sauce and steamed pumpkin. You'll work in small groups in a breezy outdoor kitchen; after four hours of chopping, grinding, stirring and frying together you'll sit down to feast as friends. *www.kinvegan cookingclass.com*

Catch the latest blockbuster

You won't find buttered popcorn at Paradiso Ubud, an all-vegan cinema, cafe and boutique. Instead, top your movie snacks with savoury nutritional yeast or sweet cacao and combine it with a homemade vegan ice cream. Afterwards, have a full dinner at the cafe or browse the locally made yoga pants and jewellery. *www.paradisoubud.com*

Previous page: The Yoga Barn is a true spiritual retreat. Left: The Ubud venue keeps guests refreshed with healthy fruit juices.

LEARNING THE LINGO

Indonesian is the most common language among workers in Bali's travel industry. Vegan in Indonesian is 'vegan'. To say 'I am vegan. I do not eat meat, seafood or dairy' it's 'Saya seorang vegan. Saya tidak makan daging, ikan, atau susu.'

Top Restaurants

SEEDS OF LIFE, UBUD

In the spiritual mecca of Ubud, this raw vegan cafe serves some of the most creative vegan fare anywhere: fermented veggie sliders, creamy-sour laksa soup with courgette noodles, beetroot 'rawvioli'. The place is also known for its elaborate health tonics with ingredients such as peony root and astragalus. *https://theseedsoflife cafe.com*

PELOTON SUPERSHOP, CANGGU

Proving that vegan food doesn't need to be healthier-than-thou, this fun Canggu spot does comfort cuisine like waffle fries, soy 'schnitzel' and brownies. *www.peloton supershop.com*

KOKOLATO, UBUD

This Ubud vegan gelato company's desserts are served in restaurants across the island in flavours such as smoky vanilla, raw cacao and turmeric-ginger-lime. *www.kokolato.com*

LOCAL CUISINE

⋙⟶ Since Balinese cuisine rarely uses dairy, most non-meat options are vegan (watch out for fish paste, though). At local restaurants called *warungs*, the food is often displayed in the open so you can pick what you want. A good bet is *nasi campur*, the classic Indonesian meal of rice with sides. Vegan options include potato cake, veggie curry and coconut-topped green beans. *Gado gado*, another beloved Indonesian dish, is a salad of steamed veg and tofu topped with creamy peanut dressing. Just skip the traditional boiled egg and prawn crackers. When it comes to street food, fried veggie fritters are ubiquitous (and delicious).

GREEN TIP

Japan's many, many convenience stores can save the day if you can't find a vegan-friendly restaurant. Onigiri (rice balls) filled with umeboshi (pickled plums) are usually a filling bet.

Practicalities

✈ Tokyo, Kyoto, Osaka, Okinawa.

💼 What you'll need depends on the season and location: winter in Sapporo can be extremely cold, while summer in Okinawa is tropical. Plan for a more conservative look in rural areas.

📅 Skiing? November–April in Hokkaido. Diving and snorkelling? July–October in Okinawa. Cherry blossom season in Tokyo and Kyoto is March and April; expect crowds. Seasonal fruit picking occurs May–June for strawberries; September–November for apples.

$ $ $

Japan

Travelling in Japan means revelling in the country's exquisite attention to detail and aesthetics. Whether it's wrapping a gift, making a cup of tea or chopping a cucumber salad, everything is done with lavish care.

Relax in the garden of a millennium-old temple complex with a cone of soy ice cream. Eat at a raw foods restaurant hidden on the 68th floor of a Tokyo skyscraper, the neon outside so bright you can't tell if it's night or day. Hike through a silent, rain-glistening pine forest, then warm up with a tofu hotpot at a rural inn. Lie on a bone-white beach, sipping from a fresh coconut whenever the tropical sun gets too hot. These are just some of the delights you can expect travelling as a vegan across the four major islands of Japan.

Japan is admittedly not always the easiest place to get by as a vegan, largely because of the language barrier and the ubiquity of fish as seasoning. But with some knowledge and advance planning, you could be eating some of the most superb plant-based dishes to ever touch your taste buds. Japan takes its produce seriously. This is, after all, the land of the $100 watermelon and the country where government guidelines recommend eating up to 13 servings of vegetables daily.

Opposite: You can't go wrong with tofu in Japan. Below: The sun sinks behind the 49m-high Yasaka Pagoda in Kyoto.

Koyasan Onsen Fukuchiin

In a 1200-year-old temple on Mt Koya, this serene *ryokan* (traditional Japanese inn) has tatami mat floors, hot springs for communal soaking and gorgeous Buddhist vegetarian breakfasts and dinners. The meticulously landscaped gardens look especially beautiful in the snow.

Vegan Minshuku Sanbiki Neko

The Australian owners of this welcoming Kyoto inn are a great resource for vegan travellers to the region. Wake up to a delicious vegan breakfast, chill in the communal area and shower with cruelty-free toiletries. Find it in a quiet, residential area of the historic Higashiyama district. *www.vegan minshuku3neko.com*

ESSENTIAL EXPERIENCES

Trek from temple to temple

On Shikoku, the smallest of Japan's four main islands, thousands of *henro* (pilgrims) in white shirts and conical hats hike between 88 ancient temples, following the legendary route of 8th-century monk Kōbō Daishi. Some of the temples offer lodging and vegan meals. The whole route takes nearly two months, but you can plan for shorter mini-pilgrimages.

Watch the morning fruit auctions

Tokyo's brand-new Toyosu Market replaces the legendary Tsukiji, most famous for its predawn tuna auctions. But like Tsukiji, Toyosu also has fruit auctions, a highly recommended window into Japan's produce-revering culture. Arise while the city sleeps to stand in the observation area watching melons, white strawberries and head-sized apples sell for sky-high prices as the auctioneer chants and stomps.

Melt your stress away at the onsen

Visiting an onsen (hot spring) is the quintessentially Japanese way to relax. These gender-segregated, clothing-free springs range from humble pools to elaborate spa complexes. The best ones are in the mountains and there's nothing like sitting in a steaming, cedar-scented tub while snow collects on your shoulders. Attached restaurants generally offer light, vegan-friendly food.

LEARNING THE LINGO

Print out a card reading '私はビーガンです。私は一切の肉、鶏肉、魚、もしくはシーフード、また乳製品、卵、ハチミツなどを含む一切の動物に由来するものを食べません', meaning 'I am a vegan. I do not eat any meat, poultry, fish or seafood or any animal products including all dairy products, eggs and honey.'

Left: Visit famed Fushimi Inari Shrine in Kyoto then dine on oyaki, noodles and tofu at Minshuku Sanbiki Neko.

T'S TANTAN, TOKYO

Slurping up a bowl of ramen is a top Japanese experience but it's rare to find pork- or fish-free broth. This casual Tokyo Station ramen shop is all vegan, serving a deeply savoury sesame-based broth and chewy noodles topped with carrot and daikon.

SHIGETSU, KYOTO

Sample traditional Japanese Buddhist cuisine at this tatami-floored restaurant on the grounds of Tenryu-ji Temple. You'll get a series of delicate small dishes such as sesame tofu, soy-braised mushrooms, miso soup and aubergine. Lunch only. *www.tenryuji.com/en/shigetsu/index.html*

GREEN EARTH, OSAKA

Eat soy burgers, pizza or Japanese-style curry at this long-standing spot, perfect for an easy-going, affordable lunch or dinner. Don't miss the soy cheesecake. *www.osaka-vegetarian-ge.com*

LOCAL CUISINE

⟫⟶ Feast on always-vegan *shojin ryori*, which is traditional Japanese Buddhist cuisine and often available at temple restaurants. Or go for sushi with cucumber, pickled plum or bamboo. Many *yakitori* (grilled skewers) joints have veggie options, just avoid the sauces. Plenty of traditional desserts are vegan and soy ice cream is widely available. The main trick to vegan eating in Japan is dodging *dashi*, the fish-based stock that flavours everything from sauces to rice to pancake batter. Also watch for dried *bonito* flakes, sprinkled atop many savoury foods. And Japanese-style bread often contains milk or eggs.

GREEN TIP

Don't assume that 'eco-tourism' means 'vegan-friendly'. In Scandinavia, eco-tours might involve fishing, husky dog-sledding and even grouse snaring, so always ask to see the full itinerary for guided tours and excursions.

Nordic nations

Pristine lakes and forests invite nature escapes, while fjords and glaciers fire the imagination. The wilds of Nordic Europe restore the weariest of souls, and there's a vegan revolution brewing in the cities.

The epic landscapes of Nordic countries stun visitors to silence: clambering Sólheimajökull glacier in Iceland, the only soundtrack is the crunching of ice under boots, while the nerve-jangling view from Norway's Pulpit Rock or the flicker of the aurora borealis on a winter night defy description. The region's extraordinary beauty encourages contemplation and vegans feel right at home with the Nordic mindset, which reveres and respects nature.

Aspects of Nordic culture can, however, be more challenging for vegans. For example, animal skins are traditional decorations in many lodges and restaurants. But more and more, Nordic citizens are omitting meat from their diet. Health and raw food cafes are common in cities, oat and plant milks accompany a sophisticated coffee scene and acclaimed 'new Nordic' restaurants increasingly cater to vegans.

Scandinavia's world-famous design scene also comes cruelty-free. Eco-consciousness is at the heart of numerous Scandinavian brands: browse vegan fashion at Gothenburg's Thrive or vegan essentials at Copenhagen's Woron. Sweden also has a dozen or so tattoo artists working with vegan inks, mostly in Malmö and Stockholm, should you want a more permanent souvenir.

Opposite: The full glory of the Northern Lights can be experienced at Iceland's Godafoss waterfall. Above: Take a sauna by the water in Finland.

Scandic Hotels

Offering plant-based breakfast options and eerily meat-like burgers, this hotel chain's provision for vegan guests is one of several eco-friendly measures. There are Scandic Hotels across the region, including in Stockholm, Oslo and Helsinki, all with crisp, contemporary rooms. *www.scandichotels.com*

Guldsmelden Hotels

This stylish hotel chain offers vegan options for breakfast and undergoes rigorous and regular sustainability audits. Rooms have a hip boutique aesthetic and there are locations in Oslo, Reykjavík, Copenhagen and Aarhus. *www.guldsmedenhotels. com*

Lilla Sverigebyn

An all-vegan resort near Vimmerby, one of Sweden's oldest towns. Cabins are spartan but the forested lakeside setting is utterly reviving. *www.sverigebyn.se*

ESSENTIAL EXPERIENCES

Fill a basket with fruit

There's a meditative rhythm to filling a basket with apples, cloudberries or blueberries. Organic cafe and farm Rosenhill lies an hour west of Stockholm by public transport; go in October for apple-picking season. In Finnish Lapland, Visit Ranua can fix you up with a berry-picking guide. Buy a head-net to guard against the voracious mosquitoes that lie in wait in berry-studded bogs. *www.rosenhill.nu*; *www.visitranua.fi*

Flex your green fingers

Swedes are wild about organic food and farm-to-table dining. At Rosendals' Garden you can get your hands dirty on a sustainable gardening course, amble between rose bushes and rows of vegetables, and snack on just-picked produce in the cafe. *www.rosendalstradgard.se*

Spend wild nights under canvas

Provided it's with a light footprint, respectful enjoyment of the land is considered a human right in Sweden, Finland and Norway – perfectly in tune with the vegan ethos. Wild camping is permitted but read up in advance about treading softly. In Iceland, if there's no campsite nearby you can pitch a tent on uncultivated ground for one night. Whether you're by a black-sand beach in southern Iceland or listening to birdsong in a Finnish forest, camping conditions are best from June to late August.

LEARNING THE LINGO

Before committing to memory 'jeg er veganer' (Norwegian), 'olen vegaani' (Finnish) and 'jag är vegan' (Swedish), relax. English is widely, often fluently, spoken and veganism is well understood. If in doubt, use informative apps such as Vegan Norway (www.vegannorway.com).

Previous page: A post-sauna swim in Finland's Lake Saimaa. Left: There are many spots to camp and self-cater across the region.

Top Restaurants

FEMTOPIA, STOCKHOLM, SWEDEN

The revolution will be accompanied by vegan cupcakes. That's the vibe at Femtopia, a hub for feminist and LGBTIQ activism with a vegan cafe. It's a welcoming place to enjoy coffee and pastries, and mingle with a forward-thinking crowd. *www.femtopia.squarespace.com*

URTEN, COPENHAGEN, DENMARK

At this cosy vegan restaurant, seasonal, earthy root vegetables and aromatic herbs drive the menu: stuffed courgettes, baked rhubarb, mushroom 'beef' and beyond. *www.urtenvegan.dk*

FARANG, HELSINKI, FINLAND

A sultry Asian fusion restaurant, Farang's evening tasting menus show off its artistic dishes. With notice, the gastronomic experience can be tailored to vegans. *www.farang.fi*

LOCAL CUISINE

⠀⠀⠀→ Sweden's vegan city highlights are Stockholm, student-packed Gothenburg and Malmö, where cafes, bakeries and restaurants prepare plant-based *smörgåsbord* (traditional buffet). Denmark's famous 'new Nordic' cuisine, often with foraged ingredients, increasingly has vegan variations. Plant-based milks feature in most Helsinki cafes and bean-and-oat staple *nyhtökaura* graces Finnish menus. Falafel is widespread and usually vegan and some hot-dog stands have tofu options. In Lapland and Iceland (outside Reykjavík), many people consider animal fats essential for surviving brutal winters and vegan options are fewer. Prices are high, even in supermarkets.

GREEN TIP

For vegan versions of cheesy Swiss specialities, you'll need to DIY. Look for Vegusto alternatives to raclette and fondue cheese. Many cheesemongers lend the melting apparatus for a fee.
www.vegusto.ch

✈ Zürich, Lausanne, Switzerland; Lake Como, Trieste, Turin, Northern Italy.

💼 Think outdoorsy but stylish: walking boots, clothing that suits yoga stretches and blustery hikes, and smart shoes. It's likely you'll self-cater for part of this trip; don't forget a corkscrew.

📅 High summer is popular, but spring and autumn have sunny days, thinner crowds and, in September, Bologna's VeganFest. Aim for weekdays in the Italian Lakes or they'll be packed with Milanese weekenders. Ski resorts are snowiest from late December to March.

$ $ $

Switzerland & Northern Italy

Switzerland is a fantasy of chocolate-box villages beneath cloud-shrouded peaks; Northern Italy seduces with lakes and ski resorts. Heightening the appeal are idyllic wineries and vegan-friendly cities.

For frazzled travellers, arriving in Switzerland instantly hits the reset button. Lausanne and Montreux, on the shores of Lake Geneva, rose to prominence as genteel 19th-century resorts; get scrubbed in a spa before strolling between their Gothic edifices and lakefront paths. Further east are graceful Lucerne and sleek Zürich, each with vegan cafes in their worldly portfolio of cuisines. Wherever you roam in Switzerland, let the views work their restorative magic: survey mountains from an infinity pool in Zermatt, clink glasses of herb-tinged liqueur in St Moritz or grab a vegan sandwich in Geneva and picnic overlooking the Jet d'Eau.

Across the Italian border and surrounded by Piedmont vineyards lies increasingly vegan-friendly Turin. Continuing east, lower your Ray-Bans around the deep, cobalt-blue Italian Lakes. At Lake Como, embark on a stretch of the scenic walking path, Sentiero del Viandante, vegan ice cream in hand. Over in Lake Maggiore, admire butterscotch-coloured villas before committing to aperitivo hour. In far-easterly Trieste, amble across Piazza Unità d'Italia to savour sea air and a slice or three of mouthwatering marinara pizza.

Opposite: The village of Varenna on the shores of Lake Como. Above: Take in the sights of laid-back Turin, including its Mole Antonelliana.

DREAM DIGS

Hotel Schwarzschmied

This South Tyrol four-star in Lana boasts light-flooded rooms, while on-site yoga classes unkink shoulders and calm addled brains. Find plant-based breakfasts and a daily organic vegan dinner menu in the restaurant, La Fucina. *www.schwarzschmied.com*

Hotel Swiss

A short stroll from the south shore of Lake Constance, eco-conscious Hotel Swiss in central Kreuzlingen is 100% vegan. At the gourmet restaurant (Fridays only), chef Raphael Lüthy prepares three- to five-course dinners paired with vegan wine. *www.hotelswiss.info*

Hotel Gabry

Next to Lake Garda, Hotel Gabry serves home-baked vegan cakes and bread, soy yoghurt and plant-based milks. Its rooms are awash with colour, plus there's a mosaic-floor pool and free spa. *www. hotelgabry.com*

ESSENTIAL EXPERIENCES

Soothe away aches and pains

These mountains make a heady backdrop for soothing, back-cracking retreats. Thermal baths such as Terme di Premia in the Italian Lakes have hydromassage jets and whirlpool tubs. Another worthwhile address is boutique spa hotel Botango in South Tyrol, with its steam bath and Finnish sauna – bonus points for the late-riser breakfasts (vegan with advance notice). For self-improvement, Queen of Retreats offers programmes to tone bodies and heighten mental clarity in an uplifting Alpine setting. *www.premiaterme.com*; *www. botango.it*; *https://queenofretreats.com*

Taste a vegan tipple

We'll spare you the gory details of isinglass, albumen and gelatin, the animal products often used to clarify wines. But rest assured, everything at Venturino Vini in Piedmont, from ruby-red grignolino to sparkling rosé, is certified as vegan. Make sure you request a tasting a couple of days in advance at this family-run operation, dating back to 1928. By road, it's 85km southeast of Turin. *www.venturinovini.com*

Indulge in lazy lake sports

There's no need to break a sweat to get out on the Italian Lakes. At Lake Como (*pictured*), tandem kayaks halve the effort, or you can borrow a sailboat. Swiss lake towns also invite slow living, such as Montreux and its pedalo rental, outdoor pools and lake beaches.

Top Restaurants

LEARNING THE LINGO

'Sono vegana' (or vegano, if you're male) is a simple statement of your vegan diet in Italian. 'Ci sono delle uova?' (Are there eggs in it?) is useful for weeding out non-vegan pasta. In French- and German-speaking Switzerland, 'vegan' is végétalien(ne) and Veganer, respectively.

MEZZALUNA, TURIN, ITALY

This restaurant has more than 25 years of chestnut gnocchi and chickpea-flour omelettes to its name. Deft versions of traditional Italian food, such as *cuculli* (fritters) and carbonara with smoked tofu, have made it a vegan temple. *www.mezzalunabio.it*

BEETNUT, ZÜRICH, SWITZERLAND

One of the best of Zürich's vegan cafes, Beetnut is green in both decor and menu. On the menu are Buddha bowls; toast with smashed avocado; fake tuna or chocolate banana; and filling smoothies. *www.beetnut.com*

WELCOME BISTROT, TRIESTE, ITALY

Mismatched furniture, carefree decor and reasonably priced vegan food all feature here. The menu ranges from beetroot hummus to *piadine* (stuffed flatbreads). *www. facebook.com/WelcomeBistrot*

Left: Vegan corn lasagne (top) and Sardinia bread rolls are both on the menu at Mezzaluna, Turin.

LOCAL CUISINE

⟫⟶ In Northern Italy, polenta is an art form and Alba's white truffles a near-spiritual experience. Garlicky, tomato-topped bruschetta, rosemary-spiked focaccia and pizza marinara are usually vegan-friendly; authentic dough should be no more than flour, yeast, water and salt (but check). Swiss food is heavy on cheese but look for *rösti* (potato fritters). They're vegan when fried in oil and when flavoured with apple, onion or herbs, rather than bacon. Vegan cafes – in the healthful acai bowl, almond-milk latte mould – are increasingly common in cities such as Lausanne, Bern and Zürich. Ethiopian and Indian restaurants are also good bets.

CULTURE

GREEN TIP

'Vegan', 'vegetarian' and even 'organic' may be used interchangeably. Pharmacies and health-food stores are handy for finding vegan-friendly staples, but you may be directed to products in the wrong category.

Practicalities

✈ Budapest, Hungary; Wrocław, Poland; Tallinn, Estonia; Bratislava, Slovakia; Vilnius, Lithuania.

💼 Sturdy shoes suit cobbled streets but bring smart clothes for nightclubs. Sporks and penknives are handy for picnics. Download a translation app to handle multiple languages.

📅 During balmy summer and colourful autumn, markets brim with strawberries (May) and later, cherries, raspberries and gooseberries. During Lent, restaurants in Serbia, Montenegro, Romania and Russia offer plant-based menus for pre-Easter fasting.

$ $

Eastern Europe

Cobblestoned towns and Soviet skylines form a thrilling jumble of styles in Eastern Europe, and regional cuisines are just as intriguing. Revel in the plant-based backlash against traditionally meaty dishes.

From medieval towns to avant-garde nightlife, Eastern Europe is a mosaic of cultures. Start in the capital cities of the Czech Republic, Slovakia and Hungary. In cosmopolitan Prague, you'll munch on vegan burgers in-between dawdling across storied Charles Bridge and admiring the Týn Church's needle-sharp Gothic towers. Bratislava is crowned with a Renaissance castle and its charming old town overflows with cafes, some of them terrific for vegans. Romantic Budapest marries baroque and art nouveau styles, and nightfall summons revellers to the city's zanily decorated 'ruin bars'.

And yes, the city has mastered vegan Hungarian goulash.

Poland's architectural and artistic treasures also deserve attention. Don't skip Kraków's Wawel Castle, Wrocław's 114m-long *Racławice Panorama* painting, and Warsaw's royal museum. Cultured and forward-thinking, these three cities each have small but thriving vegan scenes. Lviv, in Ukraine, is also a standout for vegans. Further north, farm-to-table food culture has nurtured a respectable vegan offering in Baltic capitals Vilnius and Tallinn. Eastern Europe's reasonable prices make it easy to roam widely and eat merrily without busting your budget.

Previous page: Paddling past the Taj Mahal in Agra, India. Opposite: Diners at Vegan Restoran V in Estonia. Above: V, in Tallinn, serves the likes of seitan and quinoa burrito with mango sauce.

DREAM DIGS

Falkensteiner Hotels, Slovakia & Czech Republic
These contemporary hotels in Bratislava and Prague pride themselves on a breakfast buffet brimming with plant-based goodies; they just need a little notice to cater for vegans. *www.falkensteiner.com*

Marta Guesthouse, Estonia
South of Tallinn's medieval Old Town, this homely guesthouse inhabits a traditional wooden building. It's proud to be the first vegan guesthouse in the Estonian capital and has a peaceful garden in which to relax. *www.martaguesthouse.eu*

Radharanė, Lithuania
This Lithuanian-Indian vegetarian restaurant features snug guest rooms. Awake to a vegetarian breakfast (vegan choices on request) and come back to lunch on samosas, tofu banana desserts and chai. *www.radharane.lt*

ESSENTIAL EXPERIENCES

Browse traditional markets
Deepen your understanding of local culture at produce markets. Combining history and fine food is Vilnius' Halės Turgus, in a lovely, century-old building where there's an excellent juice bar and New York-style bagel stall with vegan options. In Prague, riverside Náplavka Farmers Market showcases fruit and veg, jams and fresh flowers.

Join a cultural walking tour
Tours with I Like Veggie comprise four different food stops around the Czech Republic's many-spired capital. Itineraries can be tailored for vegans (and other dietary requirements) and along the way you'll view Prague icons like the astronomical clock. Over in Hungary's majestic capital, Vegan Tour Budapest leads hungry travellers to the best vegan food in between photo ops and historic nuggets. *www.ilikeveggie.com; www.facebook.com/vegantourbudapest*

Discover vegan-friendly nightspots
In an often meat-centric food culture, veganism represents rebellion against the status quo – so it's no surprise that edgy nightspots and veganism go hand in hand. Lviv's welcoming Om Nom Nom, complete with diverse cuisine and live music, is the best place to tap into Ukraine's vegan scene. In Slovakia's gritty yet pretty second city, Košice, bookish Halmi Place has a big drinks menu to help you wash down its vegan burgers. *www.facebook.com/omnomnomvegancafe; www.halmiplace.sk*

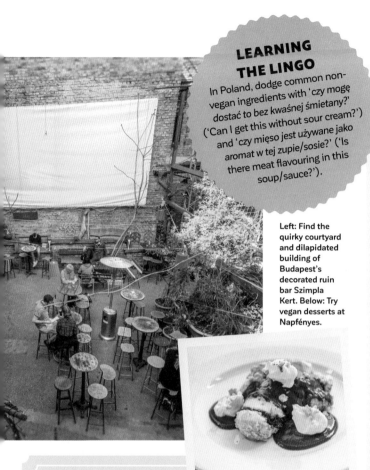

LEARNING THE LINGO

In Poland, dodge common non-vegan ingredients with 'czy mogę dostać to bez kwaśnej śmietany?' ('Can I get this without sour cream?') and 'czy mięso jest używane jako aromat w tej zupie/sosie?' ('Is there meat flavouring in this soup/sauce?').

Left: Find the quirky courtyard and dilapidated building of Budapest's decorated ruin bar Szimpla Kert. Below: Try vegan desserts at Napfényes.

LOCAL CUISINE

⫸→ Devour potato-stuffed dumplings and cabbage rolls, slurp beetroot soup and tuck into seasonal munchables such as mushrooms and berries. The use of pork 'just for flavour' and sour cream as a garnish creates a challenge but vegan-friendly restaurants have adapted signature foods with ease. Capitals and second cities have lively vegan food scenes and cafes offering crunchy falafel are pleasingly common. Markets are at their best in late spring, summer and early autumn, but year-round you'll find tempting jams and jellies, along with homemade cider, infused vodkas and, in the Balkans, bottles of *rakia* (fruit brandy). Sip slowly...

VEGA BAR WEGAŃSKI, WROCŁAW, POLAND

In vegan-friendly Wrocław, Vega Bar opened in 1987 as Poland's first veggie and vegan venue. Takes on classic Polish dishes include sour soup *żurek* (with soy-based sausage) and *gołąbki* (rice and veg-stuffed cabbage). *www.facebook.com/Vega.Bar. Weganski.Wroclaw*

NAPFÉNYES RESTAURANT, BUDAPEST, HUNGARY

Goulash with dumplings, stuffed cabbage leaves, sweet noodle cake, vanilla sponge... Classic Hungarian dishes are expertly made, while pizzas come with non-dairy cheese. *www.napfenyesetterem.hu*

VEGAN RESTORAN V, TALLINN, ESTONIA

An inventive approach to fruit and veg leads to 'faux smoked salmon' (made from carrot), quinoa-turnip cutlet and a pavlova with avocado cream. *www.vonkrahl.ee/en/vegan-restoran-v*

GREEN TIP

Tap into India's vegan scene via platforms such as Vegan First and Vegan India, and keep an eye out for vegan festivals like Mumbai's Ahimsa Fest and Bengaluru's Wilderfest. www.veganfirst.com; www.veganindia.net

Practicalities

✈ Mumbai, Delhi, Pune, Bengaluru (Bangalore), Hyderabad, Goa, Kerala, Rishikesh, Mysuru (Mysore).

💼 Light cotton fabrics are best for hot regions, cold-weather clothing for mountainous areas. Bring a head-covering scarf and slip-off shoes for holy sites, and dress to respect local culture.

📅 November or December to March is high season, with warm days and cool nights. The hot, humid April–June monsoon sees lower prices and fewer crowds. The shoulder months of April, August and October can be good for Ladakh and high Himalaya.

$ ☐ ☐

India

India is a foodie paradise for vegetarians but the vegan movement is also on the rise, making the subcontinent a tantalising adventure for vegan travellers keen to savour its cultural riches and natural wonders.

Whether amid the snow-dusted heights of the Himalaya or in the steamy palm groves of the south, India is a feast for all the senses – and the subcontinent's delightfully varied culinary terrain is a key draw for culture-hungry travellers. Throw in teeming bazaars, an awe-inspiring parade of holy sites and historical monuments, thousands of kilometres of golden beaches and protected forests prowled by tigers and elephants, not to mention some of the world's most glorious mountainscapes and trekking opportunities, it's no surprise that the homeland of yoga tops many a dream-destination list.

Thanks largely to its Hindu, Jain and Buddhist populations, vegetarian diets have long been carefully (and deliciously!) catered for in India, but veganism is also now gradually gaining ground. Some of the top destinations for plant-based travellers are India's major metropolitan cities (which are adapting to the demands of their increasingly eco-aware populations), such as Bengaluru, Mumbai, Delhi and Kochi (Cochin), as well as places that are popular for their excellent wellness and yoga scenes, including Goa, Kerala, Mysuru, Auroville and Rishikesh.

Opposite: The Mughal architecture of Delhi's Red Fort. Above: Tigers are the star attraction in Bandhavgarh National Park.

Neeleshwar Hermitage
This luxe, plastic-free eco-resort sits on an unspoilt stretch of golden beach in untourised northern Kerala, with glammed-up wood-and-thatch cottages designed according to the architectural principles of Kerala Vastu. Yoga, meditation, Ayurveda and culinary classes complement the delicious vegan-friendly cooking, which ranges from coconut-mango curry to tropical fruit bowls. *www. neeleshwarhermitage.com*

Bean Me Up
Famous for its award-winning vegan restaurant (see opposite page), earthy-toned Bean Me Up in Goa is all about healthy living and unpretentious, individually designed rooms set in gardens of coconut, mango and jackfruit trees. Dawn yoga classes run daily, while the on-site boutique revolves around organic fair-trade fashion and local artisans. *www.beanmeup.in*

ESSENTIAL EXPERIENCES

Learn to cook with an expert
Chef and food writer Anjali Pathak runs interactive cookery courses delving into local gastronomic history in a sleek kitchen in Khar, Mumbai; get in touch in advance for vegan adaptations. *www. flavourdiaries.com*

Prepare a lip-smacking feast
Uncover the secrets behind the coconut-scented cuisine of Kerala with prize-winning author/chef Nimi Sunilkumar, who delivers popular hands-on cooking classes (*pictured*), which are easily made vegan, at her Munnar home high up in the Western Ghats. *www.nimisrecipes.com*

Slow down in Bengaluru
Writer and Bengaluru expert Poornima Dasharathi heads up the team behind Unhurried Tours, which leads cultural-historical itineraries, vegan-adaptable food walks and cycling tours around India's IT capital. *www.unhurried.in*

Uncover India's stories
Much-loved Chennai-based Storytrails weaves together local history and myths in super-urban tours and a story-fuelled, vegan-adaptable South Indian 'cookathon' (called the Spice Trail). Also in Madurai, Puducherry and Trivandrum. *www. storytrails.in*

Eat your way around northern India
Reputable operator Intrepid Travel hosts an eight-day foodie extravaganza through Delhi, Agra and Jaipur, with all meals specifically vegan. *www.intrepidtravel.com*

Top Restaurants

LEARNING THE LINGO

Check whether curries are based on cream and/or ghee (clarified butter) and whether they include paneer; if so, these can usually be substituted for vegetable or coconut oil. Watch out for breads and sweets, which are often made with dairy.

Left: The self-explanatory energy bowl at Sequel, Mumbai. Below: The setting at Goa's Bean Me Up is as leafy as the menu.

BEAN ME UP, GOA

Soak up the beachy garden vibe over salads, coconut-milk smoothies, ghee-free dhal and house-made tofu curries and scrambles at Goa's favourite plant-based healthy-eating hangout. www.beanmeup.in

SEQUEL BISTRO, MUMBAI

From potato waffles with coconut yogurt to saffron-infused almond milk, organic farm-sourced fare, vegan and gluten-free, is the key to success at two Mumbai branches of Sequel. It's the brainchild of superfood specialist Vanika Choudhary. https://sequelmumbai.in

CARROTS, BENGALURU

Palak (spinach) tofu, *aloo tikki* (potato patties), *channa masala* – Indian classics get a makeover at Bengaluru's original vegan restaurant, rooted in an ethos of organic farming, seasonal ingredients and community collaboration. www.carrots-india.com

LOCAL CUISINE

➤➤➤ India's cuisine is a whirl of flavours and spices, and each region crafts its own specialities. Dairy is key and veganism is still not widely understood; many curries are ghee-based and 'pure vegetarian' is taken to mean without eggs (so including cheese, milk, etc). That said, many classic dishes become vegan by substituting or removing the dairy element; a thali, for example, without curd/yogurt. Many South Indian favourites, especially in Kerala, are based on coconut milk/oil. Other safe bets range from dhal (lentil curry), *channa masala* (chickpea curry), *aloo gobi* (cauliflower-potato curry), *panipuri* (veg-stuffed deep-fried dough) to vegetable biryani and samosas.

GREEN TIP

Finding vegan food won't be a challenge in Israel but make sure you're eating the freshest. Visit an outdoor market (or *shuk*) to see what's in season.

Israel

Appealing to history buffs and beach bums alike, Israel has national foods, hummus and falafel, which are already plant-based. And the country (especially Tel Aviv) has recently experienced a vegan boom.

The Israeli landscape packs snow-capped mountains, a desert and Mediterranean beaches into its petite frame, and its tablescape likewise encompasses an edible map of the world. *Yemenite schoog* (a coriander-based condiment) gets spooned on to breaded Viennese schnitzel, Moroccan couscous is served alongside Georgian *khinkali* dumplings, and, for dessert, try Persian *malabi* pudding, or cubed watermelon with salty Bulgarian cheese. Israel's immigrants imported familiar flavours from home, while keeping tabs on what their neighbours were having for dinner. Just as local history is a hotchpotch of religions and empires, the national kitchen merges eclectic influences, with verve. Why yes, you can have your borscht, and Turkish *bourekas*, too.

What does this mean for the vegan traveller? A trip to Israel is a chance to stamp your culinary passport with plant-based versions of traditional international fare – all for the price of a single plane ticket. When a vegan wave swept Israel a few years ago, converting roughly 5% of the country's population to a plant-based diet, locals didn't forego eating their grandmother's home-cooked specialities; they just adapted them. *Beteavon* (or, bon appétit)!

Opposite: Savour languid beach strolls at Jaffa. Above: Hard-to-beat hummus and flatbread at 416.

DREAM DIGS

Amirim

Decades before plant-based diets were trendy, this northern resort town overlooking the Sea of Galilee was founded as exclusively vegetarian (and has since catered to vegans). You can book spa treatments and simmer in a Jacuzzi near verdant landscape views. Culture lovers will find an outdoor sculpture garden and live music recitals every Friday evening during summer months. *www.amirim-home.co.il*

Abraham Hostel

This popular hostel network has a comprehensive roster of specialised tours, including an almost daily vegan tasting crawl to four Tel Aviv eateries (plus ice cream for dessert). Abraham Hostel's Jerusalem location also hosts an inexpensive hummus-making workshop and dinner four nights a week. *www. abrahamtours.com*

ESSENTIAL EXPERIENCES

Take an edible history lesson

History is the centrepiece of Delicious Israel's market tours and cooking classes. Whether you're sampling treats at a sweet shop that hasn't changed since 1935 or chatting with fourth-generation vendors in Tel Aviv's Carmel Market, these vegan-friendly tours marinate participants in the chronicles of Israeli cuisine and the neighbourhoods keeping culinary traditions alive. *www.deliciousisrael.com*

Mill around a traditional tahini factory

The basic principles for making vegan superfood, tahini, haven't changed much over the past century. The two brothers now operating the Jebrini family's 146-year-old tahini mill in East Jerusalem, near Damascus Gate, still hull, roast and grind sesame seeds on vintage basalt millstones as per their grandfather's technique. The resulting creamy spread is beloved by the likes of London-based celebrity chef Yotam Ottolenghi.

Behold Bauhaus

Pair the pearly white Bauhaus buildings of Tel Aviv's Unesco-designated architecture with the millennial pink walls of popular vegan cafe, Bana, for a design-heavy day. To do so, join the Tel Aviv Municipality's free Saturday morning English-language Bauhaus tour, and see some of the city's 4000 modernist structures. The tour ends near plant-based Bana, where reservations are recommended and the outdoor seating is delightful. *www.tel-aviv.gov.il*; *www. banatlv.com*

LEARNING THE LINGO
Most restaurants have English menus and many national chains (such as Aroma, Landwer Café, Greg Café and Joe) have vegan options. To ensure your preferences are understood, say 'ani tivoni' ('I am vegan'). Parve designates meat and dairy-free kosher food.

Previous page: Temple Mount and Dome of the Rock in Jerusalem. Left and below: Bana, Tel Aviv, is a slick spot to enjoy a chia seed pot.

LOCAL CUISINE

⇒→ Vegan food is everywhere in Israel, from casual hummus spots where the warm chickpea dish is fresh daily, to port-side fish restaurants serving meal-sized assortments of 10 to 15 different salads as the first course. Don't miss the *shakshukas* substituting tomato-poached tofu for eggs, available at most breakfast restaurants, or a *sabich pita* sandwich devoured at a street stall and packed with everything but the hard-boiled egg: fried aubergine, tahini, salads, pickles and curried mango. For snacks in between these hearty meals, pick up individually portioned sesame halva from a supermarket, or order freshly squeezed juice from a street vendor.

Top
Restaurants

SULTANA, TEL AVIV
For that roasted-meat-on-a-spit experience (minus the flesh), head to Sultana, a vegan shawarma joint. Stuff your Iraqi pita with roasted seitan or skewered wild mushrooms, plus a smattering of salads and pickles.

HUMMUS SAID, AKKO
Israelis drive long distances to Hummus Said, a veteran establishment in the port city of Akko. Don't be deterred by chipped bowls and linoleum tables – Abu Said's three hummus varieties have the hungry queuing out the door.
www.akko.org.il/en/Hummus-Said

416, TEL AVIV
This vegan hipster haven is a solid choice either for dinner or photogenic cocktails. The international-inspired menu includes plant-based versions of fried calamari, *labane* (a strained-yogurt cheese) and *fatoush* salad.
www.416.co.il

GREEN TIP

Though there's a growing number of vegan-centric shops across Spain, be sure to stock up on your usual nutritional supplements before travelling, as they may not be readily available.

Spain

Don't let the Spanish passion for jamón fool you; this sunny, fun-loving, Mediterranean-hugging nation is becoming a rewarding destination for vegan travellers, with many a delicious plant-based menú del día.

Famous for its outstanding ham, cheese and seafood, Spanish cuisine doesn't immediately scream vegan (or even vegetarian). But, contrary to the stereotypes, people in Spain eat a lot of vegetables and fruit, and the culinary scene is rapidly changing. Plant-based diets are on the increase and most major Spanish cities now have a blossoming line-up of vegan restaurants. Even in the smaller towns and villages, regional cuisines often encompass meat-free dishes (or adaptations), from cool gazpacho to crisp salads, and fresh local produce abounds. The key is in making your dietary requirements clear.

The big cities are Spain's top vegan-friendly destinations, including capital Madrid (where the party doesn't end until breakfast), Barcelona (Catalonia's jewel of Modernista architecture), Basque San Sebastián (Spain's pintxo paradise) and, down south, Granada (with its Moorish Alhambra), Málaga (the arty seaside hub of the moment) and Seville (Andalucía's soulful capital). There's also plenty of good vegan food to be found around Andalucía's wind-lashed and boho Costa de la Luz (especially Tarifa and Vejer de la Frontera), and, of course, in wellness-wild Ibiza.

Opposite: Granada's Moorish jewel, the Alhambra. Above: Carrot-lox latkas plated up at Ibiza's Wild Beets.

DREAM DIGS

Posada del Valle

This country retreat sits in Arriondas, in the foothills of northern Spain's Picos de Europa mountains. The on-site organic farm provides many ingredients for the home-cooked meals, with good vegan options on request. *www.posadadelvalle.com*

La Casa del Califa

Hidden away in the beautiful white town of Vejer de la Frontera, this rambling townhouse is one of Andalucía's most original boutique boltholes. Its creative Moroccan-Middle Eastern restaurant is a must for vegans, from its sizzling vegetable tagine to the cauliflower shawarma with rose petals. *www.califavejer.com*

Posada Magoría

In the untouristed, hiking heaven of northwestern Aragón, family-run Magoría is a rustic guesthouse specialising in lovingly prepared produce from its own organic garden, plus vegan wines. *www.posadamagoria.com*

ESSENTIAL EXPERIENCES

To market, to market

There's no better place to get under the skin of Spain's food scene than a local market – whether it's a no-frills *mercadillo* in an Andalucían mountain village or the capital's lively Mercado de San Miguel. *www.mercadodesanmiguel.es*

Sample the best of San Sebastián

Mimo specialises in cookery classes, wine tastings and pintxo-packed tours of the Basque Country's gastronomic-superstar city, with most experiences available vegan-style. Also in Mallorca and Seville. *www.mimofood.com*

Taste your way across Spain

A Taste of Spain offers a raft of food-focused tours (most of which can be turned vegan), from olive-oil-tasting trips through Córdoba's *aceituno*-carpeted hills to offbeat trails around the country's northwest. *www.atasteofspain.com*

Get culinary-creative in Barcelona

Join a trio of Galician food enthusiasts on a jaunt through Spain's culinary landscape, starting amid the stalls of Barcelona's Modernista Mercat de la Boqueria. Then you'll be back in the kitchen crafting local delicacies; ask ahead for a vegan class. *www.barcelonacooking.net*

Dance and dine in Seville

Knowledgeable guides lead you through Seville on a tapas-tastic gastro tour (vegan on request) that finishes with the foot-stomping, shawl-swishing soulfulness of a flamenco performance – olé! *www.devoursevillefoodtours.com*

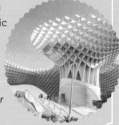

LEARNING THE LINGO

Key phrases: 'soy vegano/a' ('I am vegan') and 'no como carne, pescado, huevos o productos lácteos' ('I don't eat meat, fish, eggs or dairy'). Some dishes (such as salads or soups) can arrive topped with tuna, ham or egg; ask for them 'sin atún, jamón o huevo'.

Left: Wild Beets serves delicious breakfasts such as French toast with coconut cream.

LOCAL CUISINE

⟫⟶ Chilled gazpacho and *ajo blanco* (almond-garlic soup) and warming vegetable soups are your friends in Spain, but check chicken or fish stock (*caldo*) hasn't crept in. Vegan-friendly tapas range from *patatas bravas* (potatoes with spicy tomato sauce; watch out for mayo!) to *pimientos de Padrón* (grilled peppers), *berenjenas fritas* (fried aubergine with honey) and *espinacas con garbanzos* (spinach with chickpeas; ensure it isn't made with bone broth). Vegetable paella (check the stock), *pisto* (ratatouille) and *pan con tomate* (bread topped with tomato and olive oil) are other lifesavers.

Top Restaurants

YERBABUENA, MADRID

A vegan institution with two branches near Puerta del Sol. Find twists on potato omelettes and paella, or try smartly flavoured aubergine burgers and tofu steaks.
www.yerbabuena.ws

VEGGIE GARDEN, BARCELONA

Expertly executed South Asian-inspired cooking in El Raval, where you can feast on a rainbow of thalis, curries, pastas and burgers.
www.veggiegardengroup.com

WILD BEETS, IBIZA

Swing by the whitewashed village of Santa Gertrudis for falafel bowls, superfood smoothies and inventive original salads. *www. wildbeets.com*

HICURI, GRANADA

Dig into seitan curries, tofu tempura or vegan lasagne amid murals by street artist El Niño de las Pinturas. *www. restaurantehicuriartvegan.com*

WILDLIFE & NATURE

GREEN TIP

Brazil can be incredibly humid. Bring along electrolyte tablets or powder that you can add to water to replenish what you lose in sweat (or to help you recover after a few too many caipirinhas).

Practicalities

✈ São Paulo, Rio de Janeiro, the Amazon, Bahia.

🧳 Use a small daypack to carry items such as flip-flops, a swimsuit and sarong, sunscreen and sunglasses on hikes or beach trips. Also handy are lightweight quick-dry trousers-cum-shorts.

📅 Fresh produce is available year-round but in September and October you'll avoid major holidays, winter in the south and rainy season in the Pantanal and Amazon. Prices rise December–January and, of course, during crowded Carnaval (February/March).

$ $

Brazil

Exotic and colourful sums up Brazil's wildlife, culture and food. Markets overflow with fresh fruit, coconut products are common, pressure-cooked beans are a staple and the Amazon supplies many superfoods.

Whether snorkelling the warm waters of Ilha Grande, listening to howler monkeys from a hammock deep in the Amazon or rappelling into a cave in Bonito, nature lovers will find a playground in Brazil. The country also offers many opportunities to step back from the hustle, relax, drink a caipirinha or two and share homemade food with famously friendly locals. Even in cosmopolitan Rio, nature is never more than a quick Uber or bike ride away.

Brazilian cuisine is hearty, simple, locally sourced and, in general, fairly healthy. Bean and rice options abound and there is always a dizzying variety of fresh fruit choices. Inexpensive fresh coconut water is widely available, as are nutritious snacks such as açai bowls (just try to steer clear of the heavily sweetened ones) or Brazil nuts.

Although it's a country unfortunately known for devastating deforestation caused by cattle farming, Brazil has recently been making headlines for embracing veganism. The cities of Várzea Grande, Cuiabá and São Gonçalo have publicly committed to serving one plant-based meal every week at city-run schools, equating to five million vegan meals a year that would otherwise have contained animal products.

Previous page: An elephant may stop you in your tracks on safari in South Africa. Opposite: Atop Pico do Papagaio, Ilha Grande. Above: Jaguars inhabit Brazil's Pantanal.

Rancho dos Gnomos

Well off the beaten path, down an 8km gravel road in Joanópolis, this rustic but comfortable ranch runs an NGO dedicated to the welfare and conservation of exotic, native, domestic and migratory animals. *www. ranchodosgnomos.org.br*

Pousada Ecohar Yoga

Right on the beach in Maragogi, the 'Caribbean of Brazil', this yoga-focused centre has five apartments, yoga and cooking classes, and all vegetarian or vegan meals. *www.ecoharyoga.com.br*

Pousada Rosa Verde

Close to the gorgeous bay that Praia do Rosa is known for, this small, intimately run and pet-friendly guesthouse prides itself on being 100% vegan. *www. rosaverde.com.br*

ESSENTIAL EXPERIENCES

Find your inner gaucho

In the heart of the Brazilian savannah, ride a horse gaucho-style through the Reserva do Ibitipoca and end the night telling stories around a campfire. There are numerous restored houses for rent that are committed to sustainability and, within the reserve, you'll find Brazil's first 100% vegan lodge. *www.ibiti.com*

Wander the Amazon

Trek through virgin Amazon rainforest with your local Amazon Emotions guide, and help replace motion sensor camera cards that monitor the jaguar population in the area. Do a harnessed tree climb 60m into the canopy and catch the sunset from a canvas hammock. Watch the fireflies dance around you, interact with curious birds and end the evening lulled into relaxation by frogs and cicadas. *www. amazonemotions.com*

Spot pink dolphins

From Manaus, charter a boat with high-quality guides to see a community that protects pink dolphins. Learn about manioc flour production and taste a traditional *tapioquinha* (tapioca pancake) with coffee. Tours are not usually vegan-friendly but can be adapted given advance notice. *www.amazonecoadventures.com*

Embrace city culture

See São Paulo on a walking tour with Vegan4You. This gourmet tour starts with a vegan breakfast, then explores landmarks and vegan-centric places across the city, from São Bento station to Largo do Café and Veggie Life Store. *www.vegan4you.com.br*

Top Restaurants

Left: Chickpea and vegetable curry, quinoa with herbs, and apricot and radish salsa at Rio's .Org Bistro.

.ORG BISTRO, RIO DE JANEIRO

Rio's most passionate and cosy vegan-friendly restaurant uses whole, local, seasonal, fresh, organic and unprocessed ingredients. Conscientious chef Tati Lund is a nutritionist who studied at the Natural Gourmet Institute for Health and Culinary Arts in New York, but her food is made with more than great technique – it's made with love. www.orgbistro.com.br

POP VEGAN FOOD, SÃO PAULO

POP is the place to indulge in well-priced pizza – order the stuffed crust that's filled with locally made vegan cheese. Gluten-free and wholewheat crusts are also available. On Tuesday to Friday nights, there's an all-you-can-eat pizza rodízio, where the servers keep returning to your table with pizzas laden with different vegan toppings. www.popveganfood.com.br

LOCAL CUISINE

⋙⟶ Pay-by-weight buffets shake off their bad reputation in Brazil. They're one of the cheapest dining options available and let you fill up on brown rice, vegan *feijão* (black bean stew), okra, chickpeas, broccoli, grated carrots, and bulgur and gherkin salad. In some places, the traditional bread *pão de queijo* comes with yams not cheese, and in Minas Gerais the leaves and fruits of the *ora-pro-nóbis* shrub are used to replace meat. Brazil has a huge range of tropical fruit, and juice bars are everywhere. The local market will have a variety of fruits and vegetables year-round. Bring an empty bottle and the coconut vendor is usually happy to fill you up with fresh coconut water to go.

GREEN TIP

Veganism here overlaps with concerted efforts at environmental sustainability and a waste-reduction movement. Bring or buy reusable cloth totes and a flask for vegan coffees and smoothies.

Practicalities

✈ Vancouver, Victoria, Tofino, Sunshine Coast, Whistler, Okanagan Valley, Salt Spring Island.

💼 Mountains plus ocean means the weather can change quickly. Adequate rain gear and sturdy hiking shoes are a must. Pack for layered dressing, regardless of the time of year or exact location.

📅 June to September is high season for outdoor exploration and the best time for garden-to-table meals and bountiful farmers markets. In the Okanagan, the province's fruit bowl, fruit ripens between July and September. Vegan menus are available year-round.

$ $ $

British Columbia

'Beautiful BC' is known for its majestic mountain peaks, pristine glacial lakes, alpine flowers and lush rainforests, so it seems only natural that plant-based diets are on the up across this breathtaking province.

Birthplace of Greenpeace, environmental activism and Canada's hippie-granola culture, British Columbia boasts the highest provincial proportion of vegetarians and vegans in the country. Hippies are gradually giving way to hipsters and today veganism as a lifestyle choice is driven by a youthful and entrepreneurial spirit – call it Vegetarianism 2.0. Vancouver's Main Street seems aptly named for the movement, having spawned a handful of popular vegan-focused cafes and restaurants that no doubt inspire travellers of all persuasions to try this healthy, flavourful way of eating. Menus are often casual-chic and reflect the country's ethnic diversity, though North American mainstays such as burgers and themed 'bowls', replete with colourful veggies, grains and sauce, are ubiquitous.

The proximity of BC's cities and towns to mountains and forests, and coastal and inland waterways, also enables nature-focused travel, with vegan food soon to be as accessible as Instagrammable landscapes. Moderately priced wilderness excursions can now be enjoyed without growling tummies as more operators and outposts catch on, and they'll often modify menu offerings with a little notice (Canadians, you might have heard, can be quite nice).

Opposite: The waters of Desolation Sound prove irresistible. Above: An orca breaches in the Salish Sea.

Joe Schmucks Roadhouse Motel

A friendly stopping point near the Kootenays where travellers and hikers eat and sleep. The vegan menu quells hearty mountain appetites with safe-bet cheeseburgers and butter chick'un. *www.joeschmucks.ca*

Khutzeymateen Wilderness Lodge

Your chance to feel like a true explorer: arrive by floatplane, sail out to spot bears and sleep soundly in this environmentally and vegan-friendly (advise in advance) floating lodge in BC's only grizzly bear sanctuary. *www.khutzlodge.com*

Nita Lake Lodge

Nita Lake Lodge's on-site restaurant, Aura, hosts prix fixe Meatless Mondays and vegan selections (sourcing items from its rooftop garden). Throw in mountain and lake views and you're set for a vegan-positive slumber in Whistler. *www. nitalakelodge.com*

ESSENTIAL EXPERIENCES

Get back to nature

Spend time with goats, miniature horses and other animals during a vegan sanctuary retreat at the Arion Therapeutic Farm. Situated over 5 hectares against a mountain backdrop, it is minutes from hiking and biking trails in the Okanagan, an area known for orchards, vineyards and lavender farms. *www.ariontherapeutic.farm*

Retreat to an unspoilt island

Look out for whales and bald eagles during serene morning paddles, scout for purple starfish, or spot deer as you amble through tall cedars in the forest at the Hollyhock Retreat Centre, nestled in the wilderness on unspoilt Cortes Island. Numerous programmes include guided nature and birding walks. *www.hollyhock.ca*

Sharpen your cooking skills

Roll up your sleeves and get ready to make plant-based pastry, learn about nut cheeses and try not to get tangled in seaweed in the three-week Plant-Based Culinary Bootcamp at the Northwest Culinary Academy. You'll be based in Vancouver but can take your skills anywhere! *www.nwcav.com/ enthusiast/3-week-plant-based-culinary-boot-camp*

Pitch up by the river

Go glamping along the fast-flowing, jade-green Nahatlatch River, taking in rafting, yoga or simply enjoying the vegan food by your tent on the riverside. The power of meditation and sun salutations is amped up when you're surrounded by evergreens on a platform overlooking the waterway. *www.reorafting.com*

Previous page: Cortes Island offers stunning BC vistas. Above: In the company of animals at Arion Therapeutic Farm. Left: Break for vegan tostada at Hollyhock Retreat Centre.

THE NAAM, VANCOUVER

One of Vancouver's original vegetarian restaurants (c 1968), The Naam is trusty and affordable, open 24/7, and dishes up hearty portions, perfect after a gruelling day hike. Pick up bottled vegan sauces here, too. *www.thenaam.com*

MEET, VANCOUVER

Long queues form for MeeT's diverse vegan menu, including four spin-offs of iconic Canadian *poutine* (chips smothered in cheese and gravy), burgers and bowls, washed down with a vegan pint or two. *www. meetonmain.com*

BE LOVE, VICTORIA

Be Love restaurant is a Victoria favourite for modern plant-based and locally sourced seasonal offerings in a light and airy atmosphere. An extensive brunch menu and bowls draw travellers to make repeat visits. *www.be loverestaurant.ca*

LOCAL CUISINE

⋙⟶ Seafood is a staple of the BC diet but you can get a vegan taste of the Pacific Northwest at Fish on Fifth near the Sidney ferry terminal (Vancouver Island), West Coast Poké (Vancouver) and also trawl menus for vegan sushi. Alternatively, seek out plant-based butchery The Very Good Butchers for a rib-sticking brunch or lunch, then stock up with supplies, including vegan bangers, steak and bacon. Self-catering options abound as supermarkets and organic stores are all stocked with commercial vegan products; look for 100% vegan bakeries and speciality shops. Check out Vancouver's Vegan Resource Centre or *https:// peaces.ca* for info on seasonal markets and community dinners.

GREEN TIP

New Zealand's beloved Marmite spread is vegan and rich in B vitamins, though its salty, pungent taste is divisive.

Practicalities

✈ Auckland, Wellington, Queenstown, Christchurch, Hamilton.

💼 Seasonally appropriate casual outdoor wear will cover most of your needs.

📅 The best time to visit for both fresh produce and pleasant weather is summer, which in the southern hemisphere is December through to March, though late spring and early autumn are excellent, too.

$ $

New Zealand

New Zealand's natural splendour is truly fabulous: jagged mountains, crashing slate seas, forests so dense you feel like you've time-travelled back 1000 years, and patchwork valleys bursting with fruits and veggies.

There's a reason global billionaires are increasingly marking New Zealand as their fantasy bolthole in case of apocalypse – the islands are dreamily remote; you can drive for miles without seeing another car. Although cities like Auckland are thoroughly modern, the countryside and the wilderness feel lost in time.

Travel in New Zealand is really all about the outdoors: hiking along the golden coast of Abel Tasman while dolphins arc out of the jade waters below; watching seals play like puppies on the beach in Kaikoura; spotting cartoon-cute blue fairy penguins swooping off icebergs in glacier-carved fjords; skiing the aptly named Remarkables range; or bathing in the North Island's milky-blue thermal pools.

This natural vibe extends to the food. The country has some of the most fantastic produce in the world, from kiwi fruit and *kumara* (sweet potato) to native seasonings such as horopito and kahikatea peppercorn, and beyond to sea vegetables like karengo seaweed and bull kelp. The vegan scene is fairly well developed, especially in larger cities such as Auckland, Wellington and Christchurch. But wherever you go you'll have options, and Kiwis will usually be understanding and happy to accommodate.

Opposite: A hike in the Southern Alps is part of Aro Ha's wellness programme. Above: The kea is endemic to the South Island.

DREAM DIGS

Heritage Auckland

In an art deco former department store, this groovy hotel is home to Hectors, the first restaurant in the country to obtain Vegan Certification. An entire section of the hotel kitchen is animal product-free, so you can be worry free. *www.heritagehotels. co.nz/heritage-auckland*

Kimi Ora Eco Resort

Hike the iconic coastal track through Abel Tasman National Park, home to fur seals and little blue penguins, then rest your head at this bush lodge, with a spa and an all-veggie restaurant with plenty of vegan choices. *www.kimiora.com*

Aro Ha

This luxe wellness retreat in the Southern Alps features dawn yoga, quad-burning hikes through the snowy mountains and communal meals. The all-vegan food is exquisite, with much of the produce grown on site. *www.aro-ha.com.*

ESSENTIAL EXPERIENCES

Forage for wild veggies with a Māori chef

Near Rotorua, veteran Māori forager Charles Royal takes you on a bespoke wander through the bush, searching for *pikopiko* (young fern fronds), *horopito* (bush pepper) and *komata* (cabbage tree). You might take a muscle-relaxing soak in a natural mineral spring, ride horses through the rainy forest, or take a turn on the drums as you unwind over a cup of tea. *www.maorifood.com*

Savour local produce at the farmers markets

Relish a morning out at one of New Zealand's many farmers markets, with heaps of local produce, stalls selling hot coffee and pastries, buskers and the occasional wandering chicken. Outside Auckland, the Clevedon Village Farmers Market (*pictured*) has a bucolic charm; the subtropical Whangarei Growers Market bursts with exotic fruits; and the Wellington Harbourside has been a favourite since 1920.

Zone out in the geothermal zone

At Rotorua's WaiOra Lakeside Spa Resort, you can relax in an outdoor hot tub then work out any remaining knots with a traditional Māori massage, a geothermal mud bath or a herbal sauna. Should your belly rumble, the spa restaurant serves a vegan platter. *www.waioraresort.co.nz*

Left: The Black Forest celebration cake is one of many vegan treats at Auckland's Little Bird Kitchen.

LITTLE BIRD KITCHEN, AUCKLAND

This delightful cafe is the spot for breakfasts of sprouted buckwheat granola with nut milk, lunches of chickpea-kimchi pancakes and afternoon kombucha with cheesecake. Everything is vegan, gluten free and contains no refined sugar. *www.littlebirdorganics.co.nz*

THE LOTUS-HEART, CHRISTCHURCH

Followers of the Indian spiritual leader Sri Chinmoy run this long-time vegetarian restaurant, which has a blue-and-gold exterior resembling a temple. Almost everything on the menu has a vegan option, from grilled tempeh skewers to dosas stuffed with spiced potatoes. *www. thelotusheart.co.nz*

AUNTY MENA'S, WELLINGTON

This popular vegan cafe/ restaurant serves cheap, filling Asian dishes including wonton noodles, roti and mock meat satay.

LOCAL CUISINE

⫸→ New Zealand cuisine draws from the country's fishing and farming tradition, which means a lot of lamb and seafood but also a rainbow of fruits and vegetables, including tropical goodies such as *feijoa* (aka 'pineapple guava') and wild hibiscus blossoms. Skip British-style pubs and chippies in favour of Pacific Rim or Asian restaurants, which tend to have more veg offerings – a Thai green curry or veggie *banh mi* sandwich always goes down a treat. Hokey-pokey, a classic New Zealand crunchy toffee, is vegan. Wash everything down with beloved soft drink Lemon & Paeroa (L&P to Kiwis), so sweet it'll make your teeth chatter.

GREEN TIP

Before venturing outside major cities such as Cape Town, Durban and Johannesburg, it's wise to stock up on vegan ingredients, fresh fruit, dried fruit and nuts.

Practicalities

✈ Cape Town, Durban, Johannesburg, Garden Route, Winelands.

💼 Short sleeves and shorts are ideal for cities and safaris, though wear warmer layers on early-morning wildlife drives and in the evenings. Long sleeves and trousers help keep mosquitoes at bay.

📅 For the coastal wilds and nature of Cape Town and the Garden Route, the warm, dry months of December to February are best. To view wildlife in the country's east, parched July to October is ideal. An abundance of fruit and vegetables is available year-round.

$ $

South Africa

Follow whales and dolphins along its coastline, gaze at eagles soaring over mountains and spot safari species roaming its national parks – South Africa is perfect for celebrating the diversity of African wildlife.

Cape Town has long been heralded as one of the planet's most beautiful cities, and after you've lounged on its beaches, wandered its ever-flowering botanical gardens and looked down over the city and its dramatic Lion's Head summit from atop the wilds of Table Mountain, you certainly won't argue. What's more, you'll find that the 'Mother City' (as it's affectionately known) not only embraces its stunning natural surroundings but vegan dining, too – this cosmopolitan centre is the creative heart of animal-free eating in South Africa.

While whale watching (often from shore) on your way east through the lush coastal spectacle of the Garden Route, you'll notice that vegan options do slim considerably. However, the self-catering scene ensures that you'll have the facilities to whip up some tasty meals along the way. The same can be said of the country's famed safari parks further east, such as Kruger, Golden Gate Highlands and Marakele, where most government-owned accommodation includes access to fully equipped kitchens (private reserves tend to be all-inclusive). Picnic lunches take on a whole new meaning when you're sharing the scene with members of the Big Five (elephant, rhino, lion, leopard and buffalo).

Opposite: Luxuriate in the wild at &Beyond's Tengile River Lodge. Above: Rhino and the Big Five roam Kruger National Park.

Tintswalo Atlantic

Having risen from the ashes of Table Mountain National Park's disastrous 2015 fire, this luxurious gem (and its terrace restaurant, with vegan menu available) perches above a beach at the base of Chapman's Peak and proffers views of Hout Bay, the soaring Sentinel peak and passing whales. *www.tintswalo.com/ atlantic*

Go2Africa

This safari specialist provides tested advice to vegan travellers and can arrange your safari itinerary to take in vegan-friendly lodges. It also ensures your dietary needs are conveyed to each and every safari camp you visit. *www. go2africa.com*

Ashworth Africa

Founded by South African-born Patrick Ashworth, this specialist safari operation will create an itinerary to meet all your needs, wildlife dreams and vegan desires. *www.ashworthafrica.com*

ESSENTIAL EXPERIENCES

Learn how to un-cook vegan dishes
Raw and Roxy is popular for its raw vegan delights, which range from sweet avocado chocolate ganache and Thai curry to savoury lasagne and sushi. Join one of its Raw Vegan Un-Cooking classes so that you can bring some of this culinary magic back to your own kitchen. *www.facebook. com/rawandroxy*

Get out on safari
&Beyond's upscale, conservation-focused adventure operator offers creative, tailor-made tours and small-group safari journeys and special-interest trips. It also runs a handful of safari camps, including Tengile River Lodge (*pictured*), renowned for their service and cuisine. *www.andbeyond.com*

Eat like a local
For a plant-based adventure in the shadow of Table Mountain, join Eat Like a Local. You'll explore Cape Town on foot, soak up its unique vibe and experience up to seven vegan eating establishments. It all forms a fantastic introduction to the city, as well as its vibrant vegan scene. *www.eatlikealocal.co.za*

LEARNING THE LINGO

When English isn't spoken, try Afrikaans: 'Ek is 'n vegan' ('I'm a vegan') or 'Slegs plant-gebaseerde voedsel asseblief' ('plant-based foods only, please'). Also helpful are the following Afrikaans phrases: 'geen eier' ('no eggs'), 'geen suiwelprodukte' ('no dairy products') and 'geen diereprodukte' ('no animal products').

Left: A fynbos-based tasting platter is served during an Eat Like a Local Cape Town tour.

LOCAL CUISINE

⇛→ A traditional staple in South Africa is *mealie pap*, a stodgy maize porridge that's best served with a simple sauce of tomatoes and onions. If you can ensure the chef leaves butter and cheese out of the equation, the dish is a great vegan introduction to one of Southern Africa's most beloved dishes. When self-catering, look out for the South African brand Fry's – it produces numerous plant-based products – while it's hard to find a corner shop that doesn't sell various flavours of soya mince.

PLANT, CAPE TOWN

Cape Town comfort on a plate and no compromise needed. This long-standing vegan cafe-restaurant's salads, crepes, burgers, toasties and dim sum have put smiles on many a meat-lover's face. *www.plantcafe.co.za*

LEKKER VEGAN, CAPE TOWN

Vegan junk food is celebrated in gourmet style at Lekker's two Cape Town outlets. Sink your teeth into juicy burgers, sauce-laden wraps and sweet potato chips. Finish it all off with a chocolate 'cheeze' cake. *www.lekkervegan.co.za*

LEXI'S HEALTHY EATERY, JOHANNESBURG

Start the day in Sandton with sweet breakfast bowls and savouries such as spinach tofu scramble, before returning to lunch on dishes like black rice porcini risotto and slow-roasted cauliflower steak with chimichurri salsa. *www.lexiseatery.com*

TROPICAL

GREEN TIP

Instant noodle packets are easy to carry and can be prepared wherever there's boiling water. Some brands sold in Cambodia, imported from Thailand and Vietnam, are clearly labelled as vegan. Rice noodles are egg-free.

Practicalities

✈ Siem Reap, Phnom Penh, Sihanoukville, Battambang, Kampot.

💼 Cambodians are fairly conservative about dress. For visiting temples, bring clothing that covers your knees, shoulders and elbows. Also useful: sturdy footwear for Angkorian sites and a rain poncho.

📅 Tourist high-season runs from about November to February, with visitor numbers at their peak around Christmas and the Gregorian and Chinese New Year. Various types of mango (svay) are available all year but the most exquisite ripens from March to May.

$ ▢ ▢

Cambodia

Vegan travel in Cambodia means days spent exploring ancient Khmer temples, trekking through lush rainforests or relaxing on a palm-fringed beach, fuelled by plant-based meals and luscious tropical fruits.

The temples of Angkor are one of the glories of human civilisation. Nearby, rice paddies dotted with sugar palms surround the hydrological miracle of the Tonlé Sap lake, both filled and drained by the Mekong. Cambodia's south coast promises sublime beaches, undeveloped islands and spectacular ocean sunsets. To experience tropical nature up close, head to the rainforests of Mondulkiri or the Cardamom Mountains.

And when hunger strikes? Traditional Khmer cuisine mixes a protein with vegetables, herbs, leaves, edible flowers, spices, pickles and, of course, rice. Salads are bright and satisfyingly crunchy, thanks to green mango, lime juice, coriander, mint and lemongrass. Cambodia's indigenous Theravada Buddhism permits the consumption of animals, and even monks and nuns are allowed to eat meat if the creature has not been killed specifically for them, so an omnivorous diet is more intuitive for many Cambodians than veganism. It's possible to find all-plant versions of traditional Khmer dishes, such as *amok* (fish baked with coconut milk, lemongrass, chilli and galangal) and *samlor* (soup eaten with rice), but they're a relatively recent development, created to meet the demands of international tourists.

Previous page: Jakes Hotel, Jamaica. Left: Cambodia's famed Angkor Wat. Above: An 'Eggless Omelette' at Vibe Cafe, Phnom Penh.

DREAM DIGS

Ivy Guesthouse
A Siem Reap fixture for two decades, this backpackers' haven is a great place meet other travellers. It sports a relaxing chill-out area, great-value tapas and a restaurant-bar that's known for its extensive selection of vegan and vegetarian dishes. *www. ivy-guesthouse.com*

Four Rivers Floating Lodge
On the Tatai River in Koh Kong Province, these elegant safari-style tents, floating on pontoons, let you connect with Cambodia's tropical rainforest without sacrificing comfort. Take waterborne excursions to jungle sights during the day, then spend the evening on your private veranda amid fireflies and the spellbinding sounds of the rainforest. Staff are happy to accommodate vegan travellers. *www. ecolodges.asia*

ESSENTIAL EXPERIENCES

Pick a pepper
The world-renowned Kampot pepper boasts an internationally recognised PGI (Protected Geographical Indication), just like French wines. Organic and sustainable, La Plantation offers free tours of its 22,000-plant pepper plantation. It sells unbelievably zesty green, black, red and white peppercorns and serves Khmer and French cuisine (vegetarian options are available). Revenues help fund a local school. *www.kampotpepper.com*

Explore the Cardamom Mountains
Once a hideout for wildlife poachers, this riverside hamlet, nestled in the southern foothills of the Cardamom Mountains, has been transformed into a model of community-based ecotourism. Local guides lead visitors on forest treks, boat tours and mountain-bike adventures. Accommodation is in simple but clean guesthouses and homestays, all run by local families. The restaurant in the community visitor centre serves vegan dishes. *www.chi-phat.org*

Join an ethical wildlife-spotting tour
The acclaimed Sam Veasna Center runs one-of-a-kind tours led by knowledgeable guides, granting the opportunity to spot, in the wild, critically endangered birds and mammals, such as the blue-tailed bee-eater (*pictured*) and the yellow-cheeked crested gibbon. The organisation supports habitat conservation and local livelihoods. *www.samveasna.org*

Below: The green and pleasant interior of Vibe Cafe's Phnom Penh branch.

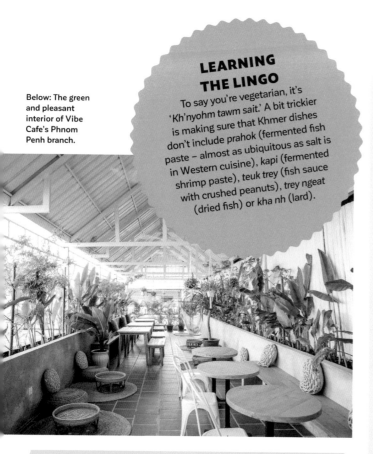

LEARNING THE LINGO

To say you're vegetarian, it's 'Kh'nyohm tawm sait.' A bit trickier is making sure that Khmer dishes don't include prahok (fermented fish paste – almost as ubiquitous as salt is in Western cuisine), kapi (fermented shrimp paste), teuk trey (fish sauce with crushed peanuts), trey ngeat (dried fish) or kha nh (lard).

Top Restaurants

MAHOB BUOS, SIEM REAP

This social enterprise serves vegan takes on Khmer classics, such as amok (made with tofu and mushrooms), red curry (with aubergine and sweet potato) and banana-flower salad. The food is delicious, the portions generous and the staff friendly.

VIBE CAFÉ, PHNOM PENH

Global favourites made with quinoa, avocado, chickpeas, tempeh and cashew cheese, plus tasty desserts, make this vegan cafe popular with travellers after wholesome, plant-based food. There's another branch in Siem Reap. www.vibecafeasia.com

VEGETARIAN FOODS RESTAURANT, BATTAMBANG

Run by an ethnic-Chinese family committed to Buddhist vegetarianism, this tiny local restaurant serves meatless offerings, all inexpensive, including soup, dumplings and homemade soy milk.

LOCAL CUISINE

⋙⟶ You'll find all-vegan cafes serving a mix of international, fusion and Cambodian dishes in Siem Reap, Phnom Penh, the south coast beaches and other spots on the travellers' map. Vegetarian restaurants are a good place to look for plant-based Khmer cuisine, as are Indian and Vietnamese restaurants, plus vegetarian cafes run by ethnic Chinese Mahayana Buddhists. At street stalls and local food shops, dishes presented as vegetarian are likely to be cooked in the same woks as non-vegan dishes and may even include seafood-based condiments. 'Vegetable soup' may be prepared with meat broth; also, fruit smoothies (teuk kalohk) may contain sweetened condensed milk.

GREEN TIP

Grand Cayman has many well-stocked shops that sell all manner of health supplements and vegan treats. If you're heading to Cayman Brac or Little Cayman, choice is limited so bring your own.

Practicalities

✈ George Town, West Bay, Bodden Town, Camana Bay.

💼 You'll likely spend most of your days on the beach, so swimsuits and light clothing are essential. Don't forget your shades and sun hat, either.

📅 High season is November–May, when it's less humid. Many avoid the Caribbean hurricane season (June–November) but the islands are quieter then and prices cheaper. February is coconut season and July is the best time for mangos and avocados.

$ $ $

Cayman Islands

Besides its white-sand coves and undersea treasures, the Cayman Islands' growing farm-to-table movement and focus on sustainability mean there's ample choice for vegans in search of a sun-drenched idyll.

Tucked away on the western fringes of the Caribbean, this trio of islands qualify as most people's idea of paradise. Palm-strewn beaches and luxury resorts are what's in store for those after some R&R, while the more adventurous will find that the Cayman Islands is also perfect for warm-water snorkelling, deep-sea diving, charged-up watersports and exploring enigmatic caves.

Grand Cayman is the largest and most cosmopolitan of the islands, with the capital George Town and surrounding coastal neighbourhoods boasting inviting restaurants and amenities catering to every kind of traveller.

And although local cuisine predominantly focuses on seafood, there's still plenty to savour if you're a vegan traveller. Tangy coconut ceviche and exotic fruits are mainstays on most menus, and enjoying your meals overlooking tranquil ocean views is of course a much-appreciated bonus.

Those keen to immerse themselves in nature will be spoilt for choice by the abundance of extraordinary tropical attractions, from idyllic coves and underwater reefs and wrecks to vegan rum-tasting sessions and green spa days.

Opposite: Azure oceans are your playground. Above: Enjoy loaded sourdough with spreads and nut cheeses at Saucha's cookery school and cafe.

Cayman Villas

If you'd prefer self-catering over all-inclusive, Cayman Villas manages dozens of properties on Grand Cayman and Cayman Brac. A personalised approach means special requests are happily accommodated, from making reservations at the best vegan-friendly restaurants to organising private chefs to create your favourite dishes. *www.caymanvillas.com*

Le Soleil d'Or

For a more secluded getaway on one of the Cayman's smaller isles, check into Le Soleil d'Or on Cayman Brac. Mediterranean-style villas are set around a lush 8-hectare farm, which supplies the resort with organic fruit and veg. Various wellness activities are on offer, from farm tasting tours to spa treatments at the beach club. *www.lesoleildor.com*

ESSENTIAL EXPERIENCES

Cook up a feast

The Conscious Café regularly runs vegan cookery classes at venues across Grand Cayman, in addition to offering a private chef service from Saucha that's ideal for those in self-catered accommodation. Meanwhile, Cookshop Bon Vivant in trendy Camana Bay also hosts cookery demos and tutorials, many of which are vegan-friendly. *www.sauchaliving.com/announcements/classes-services; www.bonvivant.ky/events-classes/adult-classes*

Shake up a rum cocktail

Learn more about the ubiquitous Caribbean rum cocktail. Grand Cayman is home to several distilleries that create the heady tipple, including Cayman Spirits Co, which offers 40-minute tasting tours. Based in George Town, the company produces a range of vegan-friendly rums, including Seven Fathoms Premium Rum, which is aged in old bourbon barrels submerged 42ft (13m) under the Caribbean Sea. *www.caymanspirits.com/tours*

Seek a spa getaway

If you want to relax with a clear conscience, there's no shortage of spas that use plant-based products. Botanika Union at the Marriott Grand Cayman turns to vegan Naturopathica products for its massages and facial treatments. Alternatively, No. 11 Spa in George Town specialises in bespoke green beauty treatments, often using locally sourced, natural ingredients. *www.marriott.comg; www.no11spa.com*

Right: VIVO's West Bay setting offers stunning sunsets. Below: Chocolate-chip pancakes with coconut vanilla ice cream at Bread and Chocolate.

Top Restaurants

VIVO, WEST BAY

VIVO is the most atmospheric meat-free venue on Grand Cayman. A rustic, candlelit setting and sunset views are complemented by farm-fresh plates such as lime-drizzled coconut ceviche topped with coconut bacon.
www.vivo.ky

BREAD AND CHOCOLATE, GEORGE TOWN

This small vegan bistro serves inventive and affordable vegan fare made from locally sourced produce. For a vegan version of jerk chicken (a Caymanian staple), tuck into an Island Bowl with grilled plantain and jerked tofu.
www.cafe.ky

ISLAND NATURALS, GRAND CAYMAN

This vegan cafe by Seven Mile Beach serves on-the-go options such as overnight oats and filled wholewheat wraps. Its shop sells health foods and green toiletries.
www.islandnaturals.ky

LOCAL CUISINE

≫⟶ Food in the Caymans is influenced by both the ocean and the spicy flavours of neighbouring Jamaica. Many restaurants have created vegan versions of traditional dishes, from jackfruit or tofu jerk to coconut calamari. Well-stocked supermarkets make self-catering easy. In George Town, Kirk Market has a range of frozen vegan foods, while the Cricket Ground near the airport hosts a farmers market. If you're staying near Bodden Town, try the Backyard Farmer for home-grown vegetables, fruit, herbs and spices. Grand Cayman also has various vegan-friendly online delivery services, including Saucha Conscious Living for artisan kombucha and Green2Go for seasonal smoothies.

GREEN TIP

Jamaica's a large island and there aren't many vegan-friendly stores outside major cities and towns. If you're not renting a car, consider bringing snacks from home to keep you sustained between meals.

Jamaica

Tropical crops and the mostly plant-based diet of the Rastafarians makes Jamaica an easy destination for vegans. With diverse dishes and a renowned laid-back attitude, it's ideal for a relaxed getaway.

Once one of the largest trading posts in the Caribbean, Jamaica is for those after sunshine, smiles and good vibes. Venture beyond the resorts and you'll find a vibrant landscape, not just in the country's natural attractions but in its welcoming cultural and culinary appeal, too.

Famed for rhythmic reggae and spicy Creole cuisine, Jamaica also boasts craggy coves giving way to verdant rainforests and magnificent waterfalls, and former plantations house farms cultivating fragrant fruits and spices. The remote Blue Mountains are criss-crossed by a number of hiking trails and fields, and are rich in coffee cultivation.

Given the right know-how, vegan travel here is a breeze. Rastafarian Ital cuisine (largely vegetarian and unprocessed), and international vegan food, is served in restaurants from Kingston to Treasure Beach, while locally grown produce is sold in both jostling town markets and tiny roadside stalls. Being on a vegan diet is more achievable in Jamaica's cities, but you'll find that rural businesses are usually happy to accommodate meat-free diets as well.

Opposite: Fresh fruit is on your doorstep at Jakes, Treasure Beach. Above: The paradisal Blue Hole in Ocho Rios.

Jakes

For those after a combination of well-being and culture, check into family-friendly Jakes at Treasure Beach. It's in St Elizabeth Parish – one of Jamaica's largest farming regions, home to the Appleton rum distillery and the YS Falls cascades. Stay in Jakes' colourful eco-conscious rooms, enjoy organically grown fare and sign up to vegan-friendly cookery classes. *www.jakeshotel.com*

Zimbali Culinary Retreats

A stunning seaside setting in Negril, Zimbali has one of the best farm-to-table restaurants in Jamaica. Stay in one of six cottages and enjoy unique gastronomic activities (also open to non-guests), including a Rasta tour that has you creating Ital dishes over an open fire. *www. zimbaliretreats.com*

ESSENTIAL EXPERIENCES

Get sensory at a Jamaican market
Kingston's Coronation Market is the largest in the English-speaking Caribbean, attracting locals, business-owners and visitors hoping for a true taste of the island. Expect stalls overflowing with seasonal produce, from ripe guavas and avocados to sweet sugar cane and sour red sorrel.

Rent a car and hit the mountains
Set off to Jamaica's legendary Blue Mountains to discover untouched scenery and quaint rural villages as well as the island's coffee plantations. Explore lush hiking trails dotted with waterfalls and book a vegan-friendly coffee-tasting tour. *www. moonjamaica.com/listing/coffee-farm-tours*

Seek out a mineral spring
In Jamaica, spas abound. For the most natural experience possible, take a trip to the Blue Hole Mineral Spring in Westmoreland. Set in a karst limestone cavern, it's a subterranean pool with refreshing mineral-rich waters, which are thought to have therapeutic properties. *www.facebook.com/ BlueHoleMineralSpringJamaica*

Learn about Rastafarian philosophy
The peaceful Rastafari Indigenous Village on the outskirts of Montego Bay is a working community that welcomes visitors to learn about Rastafarian philosophy and culture. Chat with the locals and take part in activities such as handicrafts, musical performances and preparing traditional Ital food. *www.visitjamaica.com/listing/rastafari-indigenous-village/51; www.rastavillage.com*

LEARNING THE LINGO

Though English is Jamaica's official language, words can get lost in translation as most people enjoy a meat- or fish-based diet. To be safe, tell people you only eat Ital food (typically plant-based) and no dairy, instead of just saying you're vegan.

Previous page: It's clear why Montego Bay is such a popular draw. Left: Airy oceanside accommodation at Jakes hotel.

CALABASH ITAL RESTAURANT, OCHOS RIOS

A health-food store, juice bar and vegan restaurant in one, Calabash is tucked down DaCosta Drive in Ochos Rios and is perfect for filling lunches or stocking up on organic produce. Hearty Ital food is served buffet-style, plus there's soy ice cream to relish for dessert.

MIHUNGRY WHOL'SOME FOOD, KINGSTON

If you're on a budget but want to sample Jamaican raw food, MiHungry is ideal. Found in Kingston's Market Place, it prepares affordable healthy dishes with a tropical spin – think raw vegan ackee pizzas and savoury plantain pies.
www.facebook.com/ MihungryWholSomeFood

LOCAL CUISINE

⫸⟶ Saltfish, jerk and curried meats are staples in Jamaica but you'll also find plenty of vegan-friendly ingredients. Ital cooking involves the same Creole flavours, with dishes typically made from ackee, plantain, callaloo and coconut. Some local restaurants won't consider fish a meat so double-check they understand and don't serve you something containing seafood stock. Safe bets include rice and peas, *bammy* (cassava flatbread) and spicy red pea soup. Finding fruit and veg is easy in the markets, and roadside stalls sell freshly picked fare. For more specialist vegan foods, Progressive Foods Supermarkets stock a good range of grains, pulses and dairy-free products.

GREEN TIP

Malay cooks make heavy use of a fermented shrimp paste called belacan in many veggie dishes. You can usually just ask them to skip it.

Practicalities

✈ Kuala Lumpur, Penang, Melaka, Kota Kinabalu.

💼 Bring hot-weather clothing, though it gets chilly in the highlands and mountains. Attire is fairly modest, especially in rural areas. Women should pack a headscarf for temple or mosque visits.

📅 Most of Malaysia is hot year-round. March to August is relatively dry in much of the country, while the monsoons start in September on the peninsula.

$ ☐ ☐

Malaysia

Often overlooked in favour of its flashier neighbours, Malaysia is a land of cultural, natural and culinary diversity. From its cosmopolitan coastal cities to the Borneo jungles, it's a country that rewards exploration.

Malaysia was multicultural centuries before the word existed. Indigenous tribal groups mingled with Muslim Malays, people of Chinese and Indian heritage, and Portuguese and other European traders. In today's Malaysia you can find Hindu temples down the street from vast mosque complexes and colonial-era Christian churches. Hipster coffee shops abut traditional Chinese teahouses. Women in business suits cross paths with schoolgirls in *salwar kameez* and men in plaid sarongs. This diversity extends to the cuisine, a thrilling patchwork of Chinese, Indian, Malay and European styles and influences. You'll find vegan food reasonably accessible, thanks in part to the tradition of vegetarianism among some Hindus and Buddhists.

Malaysia's tropical landscapes are the stuff of Instagram dreams: the rainforests of Borneo, the sugary sands of Langkawi, the psychedelic reefs of Pulau Perhentian, the hot sleepy market towns of Sarawak. Come to dive or snorkel, to laze on the beach, to luxe it up in spa resorts, to birdwatch in the jungle, or to view orangutans in their natural habitats. You'll leave Malaysia wondering why Thailand and Bali get all the attention – but grateful for the lack of crowds.

Opposite: Tea plantations of the Cameron Highlands. Above: The mountains make for great hikes.

Borneo Highlands Resort

Seeming to float amid the mists of the emerald Penrissen Range is this tranquil eco-resort, comprising 40 rooms and cabins and an all-vegetarian menu. Fuel up before heading for a waterfall hike, birdwatching excursion or massage. *www.borneohighlands.com.my*

The Danna

This luxury hotel on the tropical island of Langkawi has a classic colonial ambience – think black-and-white checked floors, teak furniture, a sprawling veranda. Many of the veggie items are vegan appropriate. *www.thedanna.com/en*

Banjaran Hotsprings Retreat

In a geothermal valley surrounded by rainforest, this stunner of an eco-resort is the place to soak, meditate (in a cave!) and eat tailored-for-you meals to the sound of bird calls. *www.thebanjaran.com*

ESSENTIAL EXPERIENCES

Pick strawberries and sip tea in the Cameron Highlands

When you've had it with the heat, head to the chilly, mist-draped Cameron Highlands of central Peninsular Malaysia. Coastal dwellers have been holidaying here for nearly 100 years. Pick strawberries at the local orchards, tour tea plantations and hike in the cloud forest.

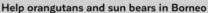

Help orangutans and sun bears in Borneo

Several top animal sanctuaries are clustered around the city of Sandakan on the northeast coast of Borneo. Watch orphaned and injured apes play at the Sepilok Orangutan Rehabilitation Centre, run by the Sabah Wildlife Department, then head to the neighbouring Bornean Sun Bear Conservation Centre, where rescued bears climb trees and lounge in the sun. *www.orangutan-appeal.org.uk/about-us/sepilok-orangutan-rehabilitation-centre*; *www.bsbcc.org.my*; *www.wildlife.sabah.gov.my*

Take a food tour of George Town

On the island of Penang, the eclectic city of George Town is Malaysia's culinary capital. Some of the best bites come from street-food vendors or stalls at open air 'hawker centres'. Enjoy veggie curries, slippery rice rolls and stir-fried greens such as *kangkong* (water spinach). For dessert, cool off with an *ais kacang*, shaved ice topped with basil seeds, sweet red bean, herbal jelly cubes, corn (*yup*) and various syrups (just ask for no condensed milk).

LEARNING THE LINGO

To say 'I am a vegan. I do not eat any meat, poultry, fish or seafood or any animal products including all dairy products, eggs and honey', it's: 'Saya seorang vegan. Saya tidak makan daging, haiwan ternakan, ikan atau makanan laut serta sebarang produk haiwan termasuk semua jenis produk tenusu, telur dan madu.'

Left: Tuck into mushroom satay sticks at Pinxin Vegan Cuisine, George Town, Penang.

Top Restaurants

DHARMA REALM GUAN YIN SAGELY MONASTERY GARDEN, KUALA LUMPUR

Admire the lotus pond and golden Buddha at this serene temple before chowing down on buffet-style vegan Chinese food in the vast cafeteria. Expect heaped plates of fried noodles, steamed buns, dumplings and fresh fruits for ultra-cheap prices. Stay for chanting with the monks.

PINXIN VEGAN CUISINE, PENANG

Go for vegan versions of Malaysian classics like tangy *asam laksa* and *nasi lemak* (coconut rice with sides) at this George Town spot in an elegant old shophouse. www. pinxinvegan.com

BEAR GARDEN, KUCHING

All the bar snacks are vegan, the back patio gets a sweet breeze, and half the proceeds go to animal welfare and conservation work. What more could you possibly want from a pub?

LOCAL CUISINE

≫→ Malaysian food brings together cuisines from across Asia and beyond. For vegans, Indian restaurants often have the most options – look for veggie-stuffed dosas, lentils and coconutty curries. Chinese Buddhist cuisine is another good bet. Malay food tends to be meaty but you can try specialities such as *nasi campur* (rice with various side dishes) with only vegetables. For dessert, *kuih*, colourful little chewy cakes made of rice flour, are usually vegan, as are many of the popular icy confections like *cendol* (shaved ice with green pandan noodles, coconut milk and palm sugar) – just watch out for condensed milk.

ROAD TRIP

GREEN TIP

Your digestion will thank you for seeking out raw food cafes in between highly processed vegan versions of local specialities. Instagram is good for research; try #vegandeutschland.

Practicalities

✈ Berlin, Munich, Germany; Salzburg, Vienna, Austria.

💼 Pack sunglasses, sat nav (or navigation app) and trail mix for an easy, snack-filled drive. Bring layers and hiking boots for the Alps and smart clothing for Berlin's bars.

📅 Summer promises beer gardens and hiking in the Alps; late September means jollity at Oktoberfest in Munich; while December's Christmas markets are fragrant with roast chestnuts and mulled wine. Ski resorts are easily reached from Munich and Salzburg.

$ $

Austria & Germany

Buckle up for cultural riches and snow-capped mountains in the heart of Europe. A road trip between the German and Austrian capitals links stately architecture, bucolic villages and cities with good vegan scenes.

Between serpentine country roads and flooring the gas pedal on the Autobahn, driving through Germany and Austria is exhilarating – and it's a breeze to zoom from picturesque mountain hamlets to buzzy, vegan-friendly cities. Begin in Berlin, where you can marvel at the mural-laden Berlin Wall, nibble in raw food cafes and dance to techno. A two-hour drive south of here is Dresden, crowned by the magnificently restored Frauenkirche; fuel up at plant-based cafes and sausage stands. Edgy art and tankards of vegan-friendly beer await in Munich, beyond which rise the Alps, providing a breathtaking backdrop to the drive on to stately Salzburg.

Capped with a fairy-tale fortress, Salzburg is an inspiring place to seek out Mozart history, re-enact scenes from *The Sound of Music*, or eat your fill in sustainable, cruelty-free cafes and restaurants. Two hours from Salzburg, stretch your legs at Stift Melk, a baroque-style Benedictine abbey perched above the Danube. From here, it's only another hour's drive to genteel Vienna. The city's imperial history and musical pedigree are accompanied by vegan schnitzels and raw cakes – a satisfying finale at journey's end.

Previous page: The Twelve Apostles limestone stacks in Australia. Opposite: A day in the sun at Berlin's Badeschiff on the Spree. Above: The region's Swing Kitchen chain excels at vegan burgers.

Hotel Flachauerhof

This tasteful, stylish contemporary biohotel in Flachau excels itself with vegan breakfast options and superfood-studded cakes. Flachau is in the middle of one of Austria's biggest ski resorts but it's equally picturesque in summer when pistes melt into flower-dotted hiking trails. Its location places you on a longer but arguably more scenic route between Salzburg and Vienna. *www. hotelflachau.at*

Almodóvar Hotel

Organic in both principles and aesthetic, this refined hotel in Berlin's arty Friedrichshain district has a vegetarian restaurant where vegan choices are always available. Come for the breakfasts of superb coffee and tofu scramble, stick around for the rooftop sauna. *www. almodovarhotel.de*

ESSENTIAL EXPERIENCES

Hit the Christmas markets

In December, towns across Germany and Austria sparkle as seasonal markets take residence in main squares. The scent of *Glühwein* (hot spiced wine) wafts across stalls, while vendors prepare skewers of fruit and stir vats of potato and onion. It can be challenging to find vegan-friendly food at Christmas events in smaller towns (and mulled wine recipes vary, so always ask whether honey has been drizzled in). Berlin has plenty of options; our favourite is Green Market, which overflows with ethical products, vegan cakes and winter warmers such as potato rösti. In Vienna, Adventgenussmarkt bei der Oper has vegan-friendly treats like sugared almonds, roast chestnuts, baked apples stuffed with nuts, and the occasional plant-based paleo cake. *www.greenmarketberlin.com; www. adventgenussmarkt.at*

Walk the vegan walk

In Berlin, the plant-based culinary scene is so diverse that different neighbourhoods demand separate tours. Follow Vegan Food Tours through edgy Friedrichshain, multicultural Neukölln or, if you're eager to hit the road, join the shorter, snack-fuelled tour of Berlin Mitte. *www.vegantoursberlin. com*

Brush up your baking skills

At its two Berlin branches, bookshop, grocer and culinary school Goldhahn und Sampson offers vegan baking classes and can customise some of its other courses for vegans. *www.goldhahn undsampson.de*

LEARNING THE LINGO

'Ich bin Veganer' is well understood. Elaborate with 'Ich esse keine Milchprodukte, Eier und Honig.' ('I don't eat any dairy products, eggs or honey.') Similarities between English and German aid in spotting non-vegan ingredients, including Milch (milk) and Käse (cheese).

Left: Getting stuck into the burgers at Swing Kitchen Rosenthaler Str, one of two outposts in Berlin.

LOCAL CUISINE

⟫⟶ Vegan travellers needn't forego sausage and schnitzels. In Dresden, find vegan *Currywurst* (spiced, saucy hot dogs) at Curry & Co outlets. In Vienna and Berlin, the Swing Kitchen chain serves vegan burgers and schnitzel sandwiches. Sauerkraut and its cold version, *Krautsalat*, is a classic Central European cabbage salad and is usually vegan. *Gröstl* (generally potatoes, onions and bacon) can be prepared vegan but, be warned, market stalls will probably have pre-mixed it with meat. Though some beers in Europe are made with animal-based finings, Germany and Austria are rightly fussy about beer purity – that means vegan brews by the barrelful.

Top Restaurants

PÊLE MÊLE, BERLIN, GERMANY

Soups, tomato and tofu ciabattas and spelt burgers fill the entirely vegan menu at Pêle Mêle, with organic beer or oat milk coffees to wash it all down. Book ahead for the excellent Sunday brunches.
www.pele-mele-berlin.de/cafe

GUSTAV, SALZBURG, AUSTRIA

This cheerful vegan cafe pours great coffee and serves sandwiches, soups and cakes. Stop by for a breakfast of fruit-laden French toast or perhaps a turmeric latte and wedge of chocolate cake.
www.gusta-v.at

SIMPLY RAW, VIENNA, AUSTRIA

Tuck into tortes at this cafe a five-minute walk from St Stephen's Cathedral. Waffles, ice cream sundaes and cupcakes grace the all-vegan menu but it's hard to look beyond the creamy tiramisu.
www.simplyrawbakery.at

GREEN TIP

Stock up on vegan supplies, from toiletries through to snacks, at The Cruelty Free Shop. It has centrally located stores in Melbourne, Sydney, Brisbane and Canberra.

East Coast Australia

Characterised by good coffee and world-class restaurants and produce, Australia's East Coast is the ultimate place to explore, stomach first, and the plant-curious are just as welcome as committed vegans.

Whether you fly or road trip between major cities, you'll find vegan-friendly communities across Australia's East Coast. From rambling Melbourne's street-art-covered laneways, admiring Sydney Opera House and tanning on a beach in sunny Queensland, to sipping your way through a wine region, it's easy to map out where to eat between activities. Checking out Luna Park and St Kilda in Melbourne? Try Matcha Mylkbar's trademarked vegan poached eggs for brunch. Bar-hopping through Sydney's trendy Surry Hills? Start with charcoal *bucatini* at spaghetti and spritz bar, Mark + Vinny's. You get the drift.

Collingwood and Fitzroy, Melbourne's coolest suburbs, are leaders in the Australian vegan dining scene, with dedicated restaurants and fast-food joints lining Smith and Brunswick Streets. Health-oriented and conscious-living movements have also prompted new venues and retreats to sprout up in various towns and cities in New South Wales and Queensland, while some places, such as the relaxed and bohemian coastal town of Byron Bay, have been welcoming plant-based diets for decades. Regardless of your destination, Australia's East Coast remains one of the very best places to travel if you're vegan.

Opposite: Drive aside the seaside on Victoria's Great Ocean Road. Above: Mock beef ragu with polenta at Melbourne's Smith & Daughters.

The Beet Retreat

Located in an iconic wine region, this Yarra Valley B&B boasts cooking classes, guided bush walks, a haven for rescued animals and shared vegan meals on a lush, 8-hectare property (moving to a new location in Queensland in the near future). It runs three-day retreats, one of which is reserved for women only. *www.thebeetretreat.com.au*

Back 2 Earth

This farm stay in Berry, New South Wales, doubles as a retreat, with raw vegan workshops, massages and a focus on sustainability. There's also a sanctuary with about 70 re-homed and rescued animals. *www.back2earth.net.au*

ESSENTIAL EXPERIENCES

Browse Australia's first permanent vegan market
Located in Miami on Queensland's Gold Coast, The Love Child is a place of cruelty-free fashion, live music, yoga classes and five as-cool-as-they-come plant-based outlets, including a burger joint, pizzeria and our favourite, a vegan ramen shop. *www.facebook.com/TheLovechild*

Treat yourself to vegan high tea
Trendy Ovolo hotel in Sydney's Woolloomooloo opened in 2018, bringing with it vegan high tea at Alibi, its plant-based restaurant. Sweet and savoury treats are served on cool, geometric stands. Sip from a choice of a dozen Rabbit Hole teas, including one that turns from blue to purple with the addition of citrus. *www.alibibar.com.au*

Sample a vegan tour
We love the philosophy behind this company, which 'takes the mystery out of veganism'. It caters especially well to the vegan-curious, exploring both Melbourne and regional Victoria. *www.veganforadaytours.com*

Scoff 'buttery' croissants
Vegan bakery Weirdoughs in Melbourne's CBD sells freakishly good pastries. The 'butter' is made from macadamia, cashew and coconut oil. Flavours fuse cuisines and creativity, from a cube-shaped croissant to others posing as a Cubano, a *banh mi* and a lobster roll. *www.weirdoughs.com.au*

Opposite page: Test the vegan food waters with Vegan For a Day Tours. Left: Macadamia milk tofu, radicchio and salted chilli at Sydney's Yellow.

Top Restaurants

SMITH & DAUGHTERS, MELBOURNE

Not your average Italian restaurant, this Brunswick St spot in Fitzroy has an all-vegan menu and punk sensibility. Secret recipes create mock beef ragu and even *trippa alla Romana* (tripe cooked Roman style) that are eerily realistic, thanks to chef Shannon Martinez's omnivorous diet. For brilliant vegan sandwiches, takeaway from nearby Smith & Deli. *www.smithanddaughters.com*

YELLOW, SYDNEY

At the start of 2016, this Potts Point hotspot switched to a 100% plant-based menu, with an impressive vegan degustation available. In the converted terrace restaurant, vegetables are treated with the care usually lavished on proteins, with heirloom varietals grown by local suppliers incorporated into dishes with everything perfectly presented. *www. yellowsydney.com.au*

LOCAL CUISINE

⫸⟶ Australian food has been influenced by immigration from countless countries. Alongside trend-driven restaurants, many traditional dishes from local Indian, Ethiopian and Middle Eastern restaurants are vegan-friendly. In major cities you'll find everything from vegan burger joints to pubs with options well beyond the dated veggie stack. Many fine-dining restaurants incorporate native Australian plants into dishes, but for a true taste of indigenous Australia, head to Torres Strait Island-owned Mabu Mabu in South Melbourne Market, which sells salads, sauces and condiments. And, in case you're wondering, Vegemite is 100% vegan – so try it!

GREEN TIP

Bring your usual backup snacks, but know that chain store Lush has 100+ outlets in the UK (80% of their eclectic range of fine-smelling products are vegan), so you're covered for emergency toiletries. https://uk.lush.com

Practicalities

✈ London, Brighton, Bristol, England; Edinburgh, Glasgow, Scotland; Cardiff, Wales; Belfast, N Ireland.

💼 For a comfortable UK adventure, you need to pack for all seasons: it could snow in spring and hail in summer! Sturdy walking shoes for cobbled streets and countryside hikes are recommended.

📅 The UK's fruit and vegetables are in abundance in late spring and through summer, so aim for May to September for fresh berries from the bushes and the best picnics.

$ $ ☐

UK

Diversity of experience is the main draw for vegan visitors to the UK. About a quarter of all the country's vegans live in London but you'll be surprised at how widespread your welcome is here.

Forget about its somewhat staid, conservative, determinedly roast-meat-and-three-veg reputation. The UK is a vibrant, multicultural place to visit, and though you'll find some of history's most important art and culture here (and, let's be honest, a few people who are sticklers for tradition), you will also marvel at the appetite for creative innovation.

Even local vegans admire the ease with which veganism has become a seemingly mandatory option at many dining establishments and transport hubs. And not just in the capital cities of London, Edinburgh, Cardiff and Belfast,

where new vegan establishments are opening in rapid-fire fashion, but right across the UK. Go to a quaint pub in Bristol for a vegan tasting menu, have 'fish' and chips by the sea in Brighton, visit a vegan fairy festival in Cornwall, enjoy vegan sorbetto in Leicester or relax in a vegan cafe with homeless cats in Sheffield. Self-catering travellers will find dedicated vegan sections in all major supermarkets and ever-expanding options in major takeaway chains. Pack your bags.

Opposite: Pull up a deckchair in Brighton, one of the UK's top vegan towns. Above: Golden beetroot dauphinois at London's Gauthier.

Hilton London Bankside
Book into the world's first vegan hotel suite, designed in consultation with The Vegan Society: all materials in the room are 100% animal-friendly. The room service is also completely vegan. *www.vegansociety.com*

The Cosy Vegan B&B
This B&B in Scotland's Fife is indeed cosy and welcomes canine visitors as well as humans. *www.thecosyvegan.co.uk*

Ambleside Manor
This glorious country guesthouse in the Lake District has luxurious rooms on two acres of private grounds. It's vegetarian but happy to offer vegan breakfast options. *www.ambleside-manor.co.uk*

The Old Rectory
A Swiss-trained chef co-owns this Victorian house in leafy South Belfast. There are vegan options for breakfast and a focus on local and organically produced food. *www.anoldrectory.co.uk*

ESSENTIAL EXPERIENCES

Create culinary feasts in the New Forest
Try 100% plant-based cooking classes on an organic farm in Hampshire's beautiful New Forest. It's not just about being resourceful and creative, there's a focus on technique. Choose three or four-hour classes, from baking to Chinese takeaway! *www.offbeetfood.com*

Sign up for a vegan retreat
UK company Neal's Yard Holidays specialises in holistic breaks, offering vegan retreats in various locations across the UK, with either a purely dietary focus or incorporating yoga and mindfulness. *www.nealsyardholidays.com/Activity/view/vegan-retreats-uk*

Entertain your inner chef
Want to learn how to make your own cheese? What about charcuterie? Street food? Beetroot Sauvage vegan cookery workshops in Edinburgh provide a surprising variety of options for curious wannabe chefs. *www.beetrootsauvage.co.uk/cookery-workshop*

Escape to the Welsh countryside
Penmeiddyn, a five-bedroom farmhouse along the Pembrokeshire Coast National Park in Wales, is the ultimate escape for inner serenity. Organic vegan cuisine awaits along with a choice of art workshops or retreats – silent, Buddhist or meditation. *www.penmeiddyn.org.uk*

Previous page: Night falls on Bristol's Clifton Suspension Bridge. Left and below: Hilton London Bankside has a suite where all foods, and even the furnishings, are entirely vegan.

GAUTHIER, LONDON, ENGLAND

London is home to the UK's first Michelin-starred venue with a vegan tasting menu: French restaurant Gauthier. Chef Alexis Gauthier turned vegan in 2015 and converted the vegetable tasting menu to 'Vegan Menu: Les Plantes'. It's been a fixture ever since. *www.gauthiersoho.co.uk*

387 ORMEAU ROAD, BELFAST, N IRELAND

A fully vegan cafe with delicious food, from breakfast burritos to 'sausage' rolls, 387 is open every day. It also has a little grocery store for all those essentials. *www.facebook. com/387ormeauroad*

VOLTAIRE, BANGOR, WALES

In Wales, head for Voltaire (25 Garth Rd, Bangor) for a vast, all-vegan menu with some inspiring and inventive choices (jerk cauliflower wings, anyone?).

LOCAL CUISINE

⟫⟶ Got a hankering for eating what the locals do? It's not so difficult. Traditional UK breakfast spreads that also happen to be vegan include Marmite and marmalade. You can source vegan black pudding from The Real Lancashire Black Pudding Company, dip a Hobnob biscuit into your tea and also partake of the locals' penchant for crisps: the Walkers brand has a range of flavours that happen to be vegan (even 'Sausage and Brown Sauce'). Drinking in the pub? Vegan cider options are plentiful: choose from Westons, Thatchers, Aspall, Merrydown or Brothers. Even good-old Guinness is vegan. All the main UK supermarkets also clearly label their vegan wines.

GREEN TIP

For road munchies, any West Coast health-food store, as well as chains like Whole Foods and Trader Joe's, will stock an enormous variety of vegan snack foods.

Practicalities

✈ Seattle, Portland, San Francisco Bay Area, Los Angeles.

🧳 Pack fleeces and raincoats for the chilly, drizzly Pacific Northwest, hiking clothes for the national parks and swimsuits for the SoCal beaches.

📅 The Pacific Northwest is far and away more pleasant in summer, as is northern California. Go then for the best produce and farmers markets. Southern California is gorgeous year-round.

$ $ $

West Coast USA

From the rain-slick streets of Seattle and the stunning seaside curves of Hwy 1 to the freeways of Los Angeles, the West Coast is built for road trips, with some of the world's best vegan food everywhere you stop.

The West Coast is the edge of America, both geographically and culturally. Cool happens here first and spreads east: hippiedom, grunge rock, the sustainability movement, coffee culture. Its wellness-minded eco-consciousness puts it front and centre when it comes to vegan cuisine. There's perhaps no better place to follow a plant-based diet than the 'Left Coast'.

So, vegans, start your engines. A road trip from Washington State to Southern California (or vice versa) takes you through some of America's most glorious landscapes and delicious vegan restaurants. Devour 'fish' tacos on the beaches of San Diego, before hopping in your car, sand between your toes. Cruise into Los Angeles to celeb-spot while dining on plant-based sushi. Follow Hwy 1 up the edge of the crashing Pacific, stopping in quaint little hippie towns for grain bowls and green juices. Park in San Francisco for vegan Vietnamese. Heading into Oregon, drive through the drizzle into Portland to feast on mock meat barbecue and check out the indie bookstores and music clubs. Then cap off your long trip with a triple scoop of vegan ice cream in Seattle.

Opposite: Sandy coves line California's beautiful Big Sur. Above: Yoga with a view in Santa Barbara.

The Stanford Inn by the Sea

This rustic-luxe resort (*pictured*) sits on a hill overlooking Mendocino Bay. Canoe the Big River Estuary, have a massage with botanical oils, or dine at the all-vegan restaurant, where the strudel stuffed with local sea palm is a must-try. The chef offers vegan cooking classes, so you can learn how to make it too. *www.stanfordinn.com*

Velo Bed and Breakfast

Just off Interstate 5 in the university town of Eugene, Oregon, Velo is a cosy little B&B on a wooded site, where the chef-owner serves vegan and vegetarian breakfasts. It's a popular spot with cyclists and those touring the nearby vineyards. *www.velobandb.com*

ESSENTIAL EXPERIENCES

Take a DIY vegan doughnut tour

Can anything beat a doughnut as a road-trip snack? The West Coast has so many vegan and vegan-friendly spots you could stop at a different one every time your tummy rumbles. Start with a chocolate-glazed pastry at Donut Farm in LA and finish with an apple fritter at Mighty-O Donuts in Seattle. *www.vegandonut.farm*; *www.mightyo.com*

Breathe deeply at Tassajara Zen Mountain Center

In summer, this iconic California Zen centre (*pictured*) opens its doors to visitors, who meditate, soak in the Japanese-style springs, and enjoy the renowned vegetarian cuisine, much of it vegan. *www.sfzc.org/tassajara*

Sip pinot at a vegan winery

From Napa to Oregon's Willamette Valley to southeastern Washington, the West has some of America's primo wine-growing territory, so designate a driver ASAP. A growing number of wineries produce products free from fining agents such as egg white or fish bladder protein. The website www.barnivore.com is a good guide to what's vegan-safe. *www.barnivore.com/wine*

Nibble away your Saturday morning in Portland

Veggie-lovers delight in the city's enormous, wildly varied farmers market, its stalls heaped with locally grown delicacies such as Willamette Valley hazelnuts and organic marionberries, as well as foraged plants like nettles and fiddlehead ferns. *www.portlandsaturdaymarket.com*

Left: Portland's thriving vegan scene is well represented in the city's bountiful farmers markets.

CROSSROADS KITCHEN, LOS ANGELES

This stylish I-can't-believe-it's-vegan restaurant is a favourite of movie stars and mortals alike. Dine on lasagne with almond ricotta or fried 'chicken' and waffles, or hang out for cocktails and late-night Buffalo-style *maitake* mushrooms or vegan sliders. www.crossroadskitchen.com

GRACIAS MADRES, LOS ANGELES/SAN FRANCISCO

Who can resist plant-based Mexican food? Try a *chimichanga* with tempeh chorizo or a squash-stuffed *quesadilla* with cashew *crema*. Seasonal produce is from the restaurant's organic farm. www.gracias-madre.com

FRANKIE AND JO'S, SEATTLE

Lick all-vegan ice cream in flavours like chocolate tahini or ginger with fresh turmeric. There's one shop in Capitol Hill, one in Ballard. *www.frankieandjos.com*

LOCAL CUISINE

≫→ Vegan food on the West Coast has no boundaries. You can get vegan versions of almost any cuisine: Thai, Ethiopian, Cuban, soul food and way more. American classics, vegan-style, include barbecued 'meat' made with tempeh or seitan, veggie burgers with all the fixings, ketchup-smothered breakfast hash browns and triple-scoop cones of vegan ice cream. Try Mexican dishes such as burritos and enchiladas, stuffed with beans and veg – just hold the cheese, please. Pasta with red sauce is a good bet at any Italian restaurant. Or go for old-school California hippie cuisine like brown rice platters and sprout-y avocado sandwiches.

ABOUT THE AUTHORS

KAREN CHERNICK

Arts and culture writer Karen Chernick pleads guilty to making restaurant reservations at a new destination first, and booking flights second. A vegetarian from age seven and now a vegan, Karen's delight in tasting new food has led her to meals ranging from dinner at a Hare Krishna temple in Prague to cooking lessons on an organic farm outside Chiang Mai, Thailand. She has learned to be adventurous, but always pack snacks, and hopes this book helps vegan travellers discover that the world truly is their plant-based oyster. When not sampling global eats, she can be found in her hometown of Tel Aviv where she writes for a range of outlets found on her website, karenchernick.com.

CATHY BROWN

High in the Andes of Argentine Patagonia, Cathy Brown lives on a small farm with her teenage children, where she plays with natural construction, organic gardening and medicinal plants. Her passion is helping to preserve indigenous cultures and she works closely with communities in the Brazilian Amazon.

JEMIMA FORBES

Jemima Forbes is a full-time writer and part-time explorer who has created guides for multiple travel publications. She took her first flight at six weeks old and has since called seven destinations home, including the Seychelles and Shanghai. Jemima now splits her time between the UK and the Cayman Islands.

DINAH GARDNER

Dinah Gardner is a writer and researcher and has been based in East Asia for more than two decades. In her humble opinion, Taiwan is the friendliest and most marvellous country she has lived in by far. Dinah is also an author for the Lonely Planet Taiwan and Taipei guides.

ANITA ISALSKA

Anita Isalska is a British travel journalist and freelance editor based in California. She specialises in Central and Eastern Europe, offbeat destinations and gourmet travel – with a focus on travelling on a vegetarian, vegan or gluten-free diet. Read her words at www.anitaisalska.com.

SOFIA LEVIN

Seasoned traveller and food journalist Sofia Levin believes that eating overseas is the best way to understand another culture. From her base in Melbourne she writes for newspapers and travel magazines, and co-authors international guidebooks. Follow her culinary adventures and ask her how to #EatCuriously on Instagram: @sofiaklevin.

EMILY MATCHAR

Emily Matchar writes about food, travel, culture, science and more for magazines and newspapers such as the *New York Times*, the *Washington Post*, *Smithsonian* and *Outside*, and has contributed to many Lonely Planet books. She lives in Hong Kong and Pittsboro, North Carolina with her family.

ISABELLA NOBLE

English-Australian on paper but Spanish at heart, travel journalist Isabella Noble is a Spain specialist but also writes extensively about India, Thailand, the UK and beyond for Lonely Planet and other publications. She's a lifelong vegetarian, busy seeking out plant-based deliciousness around the globe. Find Isabella on Twitter and Instagram @isabellamnoble.

KARYN NOBLE

Karyn Noble is an award-winning freelance writer, food judge and restaurant critic based in London. She follows a mostly flexitarian diet and appreciates the health benefits of plant-based eating. Follow her tweets @MsKarynNoble.

MATT PHILLIPS

Matt Phillips loves his food, his exercise and his travels, and always does his utmost to combine the best of them all. His most recent foray involved a solo 21km beach run along Scotland's stunning Moray Coast, before refuelling on sumptuous vegan brownies at The Bakehouse Cafe in Findhorn.

DANIEL ROBINSON

Daniel Robinson has been writing about Southeast Asia and its vegan cuisines since 1989, when he researched Lonely Planet's first, award-winning guides to Vietnam and Cambodia (co-authored with Tony Wheeler). Since then, his writing has appeared in scores of guidebooks as well as the *New York Times*, the *Los Angeles Times* and *National Geographic Traveler*.

RACHEL SAFMAN

Since first planting roots in a Thai rice paddy in the 1980s, Rachel Safman often returns to Southeast Asia. She lived and worked in the region as a sociologist for more than a decade before shifting gears and becoming a rabbi in Connecticut – but she still leads kosher tours of Southeast Asia.

CAROLINE VELDHUIS

Caroline Veldhuis is a lifelong vegetarian who started a recipe collection at the age of seven and is fulfilling her childhood dream of being a detective by sniffing out some of the best food, drinks and nature-related sights across the globe. She has contributed to several Lonely Planet Trade & Reference titles.

Published in December 2019 by Lonely Planet Global Limited
CRN 554153
www.lonelyplanet.com
ISBN 978 1 7886 8758 4
© Lonely Planet 2019
Printed in China
10 9 8 7 6 5 4 3 2 1

Publishing Director Piers Pickard
Associate Publisher Robin Barton
Commissioning Editor Christina Webb
Editors Nick Mee, Monica Woods, Lucy Doncaster
Art Director Daniel Di Paolo
Picture researcher Ceri James
Print Production Nigel Longuet
Cover image © Niki Fisher

STAY IN TOUCH lonelyplanet.com/contact

AUSTRALIA The Malt Store, Level 3, 551 Swanston St,
Carlton, Victoria 3053 T: 03 8379 8000

IRELAND Digital Depot, Roe Lane (off Thomas St),
Digital Hub, Dublin 8, D08 TCV4

USA Suite 208, 155 Filbert Street, Oakland, CA94607
T: 510 250 6400

UNITED KINGDOM 240 Blackfriars Rd, London SE1 8NW
T: 020 3771 5100

Paper in this book is certified against the Forest Stewardship Council™
standards. FSC™ promotes environmentally responsible, socially beneficial
and economically viable management of the world's forests.